Virtual Automation Environments

MATHEMATICAL METHODS IN TECHNOLOGY

Editor

Freddy Van Oystaeyen

University of Antwerp
Antwerp/Wilrijk, Belgium

1. Virtual Automation Environments: Design • Modeling • Visualization • Simulation, *Herwig Mayr*

ADDITIONAL VOLUMES IN PREPARATION

Virtual Automation Environments

Design • Modeling • Visualization • Simulation

Herwig Mayr

Upper Austrian Polytechnic University
Hagenberg, Austria

MARCEL DEKKER, INC. NEW YORK • BASEL

ISBN: 0-8247-0736-2

This book is printed on acid-free paper.

Headquarters
Marcel Dekker, Inc.
270 Madison Avenue, New York, NY 10016
tel: 212-696-9000; fax: 212-685-4540

Eastern Hemisphere Distribution
Marcel Dekker AG
Hutgasse 4, Postfach 812, CH-4001 Basel, Switzerland
tel: 41-61-261-8482; fax: 41-61-261-8896

World Wide Web
http://www.dekker.com

The publisher offers discounts on this book when ordered in bulk quantities. For more information, write to Special Sales/Professional Marketing at the headquarters address above.

Current printing (last digit):
10 9 8 7 6 5 4 3 2 1

PRINTED IN THE UNITED STATES OF AMERICA

In a time of drastic change, it is the learners who inherit the future. The learned find themselves equipped to live in a world that no longer exists.

E. Hoffer

Preface

About this Book

Modern manufacturing plants have evolved from classical "manu"-factures with isolated automated manufacturing cells into integrated production environments. In the same way, systems for design and modeling of manufacturing components, such as robots or numerically controlled (NC) machines, have evolved from CAD/CAM extensions, through robot modeling and simulation systems, into modern integrated systems with trendy names such as "virtual factory" modelers and simulators or "virtual automation environments".

This book presents the underlying concepts for design and modeling of geometry and kinematics together with the necessary visualization and simulation aspects. These concepts have remained much the same during all the evolution steps; only the complexity and capabilities have been extended considerably.

Organization of this Book

This book has been divided into four parts:

I. Design

II. Modeling

III. Visualization

IV. Simulation

thus covering, in the order in which they occur in practice, all major areas necessary for creating a virtual environment. Each part is subdivided into several chapters, discussing the key issues and focusing on their relevance for virtual environments. Most chapters conclude with an example taken from the author's personal experience using the given concepts in industrial applications. A detailed table of contents together with a comprehensive index additionally support easy access to the contents of this book.

Part I: Design

Chapter 1 gives an introduction to the rapidly growing field of virtual reality (VR). Since virtual reality is still changing from day to day, only the basic concepts and methodologies are presented.

Chapter 2 describes the origins of graphic-data processing and its evolution into computer-aided design (CAD) and computer-aided manufacturing (CAM). Additionally, the coupling of CAD and CAM into computer-supported production is discussed.

Chapter 3 covers trends in factory automation and computer-integrated manufacturing. The development of industrial robots and NC machines is sketched together with modern programming concepts to achieve a higher degree of automation and less production error.

Chapter 4 introduces the mathematical background for positioning and visualizing objects. The analytical geometry approach is pursued using homogeneous coordinates and transformation matrices.

Part II: Modeling

The representation of 3D objects is discussed in *Chapter 5*, together with appropriate definitions and several alternatives for a classification of the different representations. This chapter forms a structured basis for *Chapter 6*, which gives a detailed survey of different geometric models in 2D and 3D. Mathematical problems of solid modeling are discussed and their solutions are pointed out for the different models.

The goals of *Chapter 7* are to familiarize readers with forward (direct) kinematics and inverse (backward) kinematics of a mechanism, enable them to determine its Denavit-Hartenberg parameters, and understand the process for determining the kinematics model of a mechanism.

Chapter 8 states and classifies the problems arising from inaccurate and/or inconsistent data. Based on this classification, strategies for solving and avoiding these problems are presented.

Part III: Visualization

In *Chapter 9*, sample algorithms are presented that prepare geometric data in such a way that they can be visualized on modern, window-oriented raster-graphic displays. The importance of time-efficient algorithms is explicitly stressed.

Chapter 10 gives a survey on improving images to look more "realistic" to the user. Techniques to achieve this goal include hidden line/ hidden surface removal, rendering methods, and texturing.

Part IV: Simulation

Chapter 11 introduces the concept of simulating a system, focusing on graphic simulation. Various models for and aims of simulation are discussed. Based on simulation techniques, the verification of tasks and products is sketched together with the possibility of generating robot/ NC code automatically.

Chapter 12 gives a survey of algorithmic manufacturing verification, particularly the determination of collisions in a manufacturing cell. Several collision checking problems are stated and a survey of solutions to these problems is presented.

Chapter 13 covers the transition from conventional simulation systems to (distributed) virtual automation environments. Suitable system architectures and information flow models (e. g. software sensors) are discussed, and the concept of a virtual factory is introduced.

Examples from various areas conclude this chapter as well as the whole book and illustrate the breadth of the impact of virtual automation environments.

Intended Audiences

This book is intended mainly as a textbook giving an introduction and a survey of virtual reality techniques used in industrial automation at the turn of the second millennium. It should serve as a primer covering the interests of a technician who wants to acquire knowledge on virtual environments used in automation, and as a textbook for teachers and students at the graduate level, as well as for self-study.

For Technicians Who Want to Deepen Their Knowledge of Virtual Environments

Part I of the book gives a rather brief and basic survey of the history and key notions of virtual reality, CAD/CAM and analytical geometry. Only the basic understanding of how to think as a technical person is assumed. Depending on the background of the reader, some or all chapters of this part may be skipped.

Part II should be read thoroughly and in the order presented, first because the geometry and kinematics models introduced in Chapters 5 – 7 are fundamental for the rest of the book, and, second, because the problems of data accuracy and consistency discussed in Chapter 8 are frequently underestimated or neglected altogether, but turn out to be crucial in most real-life applications. The reader who discovers in studying Chapters 6 or 7 that he or she is not familiar enough with the concepts of transformation matrices can return to Chapter 4 for more in-depth information.

Part III (on visualization) is only loosely connected with the other parts and can be read separately as need arises.

Part IV should be read in the order presented since the transition from simulation systems to virtual environments and its consequences for modeling are discussed in these chapters.

For Teachers Giving a Graduate Course on Virtual Environments

The book has been developed and used as a textbook for a 2-semester 2-hours graduate course for technical students (i. e., a course comprising 30 hours), the students' interests ranging from applied mathematics, through computer science and mechatronics, to theoretical physics. The students, however, usually were novices in the field of virtual reality in general and virtual automation environments and their industrial applications in particular. For a two-semester course, it is suggested to put Parts I and II into the first semester and Parts III and IV into the second semester, with a brief repetition of Chapters 4, 6, and 7 at the beginning of the second semester.

For a 2-hour, one-semester course, a suggested structure could include Chapter 1 (including the key terms of Chapters 2 and 3; 1 week), Chapter 4 (1 week), Chapters 5 and 6 (3 weeks), Chapters 7, 8, and 11 (2 weeks each), Chapter 13 (2 weeks + 2 weeks for detailed examples). A one-semester course structured this way was given several times by the author at various universities throughout Europe in the frame of the EU ERASMUS project "Mathematical Methods in Technology", organized by the University of Antwerp, Belgium.

For Students Acquiring Knowledge of Virtual Environments by Self-Study

Similarly to the hints given for technicians above, only those chapters of Part I should be read that the student is not familiar with. Only Chapter 4 is required to be thoroughly understood before the reader goes on to Part II. The second part should be read in the order presented.

To make the most out of Part IV, students are advised also to study Part III, particularly the algorithms presented therein. For the reader thus equipped, the special needs of virtual environments are better understood and the examples of Part IV are easier to follow.

Feedback

The author would be very grateful for any comments on the substance of the book as well as the organization of its contents. Please reply to me either electronically at *Herwig.Mayr@fhs-hagenberg.ac.at* or at my surface mail address:

Herwig Mayr
Dept. of Software Engineering
Upper Austrian Polytechnic University
A-4232 Hagenberg
Austria

Please indicate the purpose for which you use the book and the alterations that might help you make better use of this book. Thank you!

Acknowledgments

A course on what would nowadays be called "virtual automation environments" has been given by the author at the Johannes Kepler University Linz, Austria, annually since 1988. Prof. Dr. Fred Van Oystaeyen and Prof. Dr. Rudi Penne in the Department of Mathematics and Computer Science at the University of Antwerp, Belgium, invited me to give this course in the framework of the EU ERASMUS project "Mathematical Methods in Technology" in Antwerp, Groningen, The Netherlands, and Almeria, Spain, throughout the 1990's. Prof. Van Oystaeyen and Prof. Penne also suggested turning the course notes into a book. Ms. Jennifer Paizzi and Ms. Helen Paisner of Marcel Dekker, Inc., continued the excellent guidance during the publication process.

The areas of CAD/CAM, computer graphics, and virtual reality are among those in the larger framework of computer science that change most rapidly. Therefore it took several approaches until one effort was successful to finish the manuscript quickly enough so that the information is not completely outdated before the book goes into print. I would like to thank all my colleagues at the University of Antwerp for their patient wait for the final manuscript.

Since a university professor has a rather unbalanced workload throughout the year, most of the work could be done only during "holiday time". The dean at the Upper Austrian Polytechnic University at Hagenberg, Austria, Prof. Dr. Witold Jacak, frequently encouraged me to work on this project, as did Prof. Dr. Bruno Buchberger at the Johannes Kepler University Linz, Austria. Thank you both!

Some of the basic work on this book was done by some former students of mine. I would therefore like to thank all the members of the former GeM group at RISC-Linz. In particular, Dipl.-Ing. Monika Marko, Dipl.-Ing. Leopold Peneder, and Dipl.-Ing. Werner Wolf are to be thanked for all their implementation work on the examples given in Chapter 13. Dipl.-Ing. Karl Artmann and Dipl.-Ing. Andreas Hametner did the excellent design and implementation work for the project presented in Section 8.4. The example given in Section 11.8 was prepared by Dipl.-Ing. Leopold Peneder.

The project ViFE, parts of which are presented in Chapter 13, was initiated and coordinated by Prof. Dr. Arto Toppinen of the Walter Ahlström Institute of Technology, Pohjois-Savo Polytechnic, Varkaus, Finland, and funded by the European Commission in the framework of the EU LEONARDO DA VINCI programme. I would like to thank Prof. Toppinen for all his initiative and efforts. Dipl.-Ing. (FH) Gerald Kern and Dipl.-Ing. (FH) Jürgen Pulkrab are to be thanked for implementing the first prototype of the VIFTOO-SCARA integration in connection with their diploma theses at the Upper Austrian Polytechnic University.

It is difficult to spare the time necessary for writing books when one has a family, even a most understanding one like mine. However, I consider myself an extraordinarily lucky person, since my wife, Barbara, not only accepted that I spent many nights and weekends at work, but is also the best proofreader I know. Thank you, Barbara, for all the time you took to get my sentences right. Also, a big hug to my children, Bernhard and Marlene, for missing their daddy but not dismissing him when I did not see them for days.

Thank you all!

Herwig Mayr

Contents

List of Figures

List of Tables

Part I

Design

Chapter 1

Virtual Reality

This chapter gives an introduction to the rapidly increasing field of virtual reality (VR). Since virtual reality is a field that is still changing from day to day, only the basic concepts and methodologies are presented.

1.1 What is Virtual Reality?

The term *virtual reality* is used by many different people with many meanings. There are some people to whom virtual reality is a specific collection of technologies, that is a head mounted display, glove input device and audio. Other people stretch the term to include conventional books, movies or pure fantasy and imagination. The most concise definition of virtual reality might be: "Virtual Reality is a way for humans to visualize, manipulate and interact with computers and extremely complex data." [Aukstakalnis and Blatner, 1992]

The visualization part refers to the computer generating visual, auditory or other sensual outputs to the user of a world within the computer. This world may be a CAD model, a scientific simulation, or a view into a database. The user can interact with the world and directly manipulate objects within the world. Some worlds are animated by other processes, perhaps physical simulations, or simple animation scripts. Interaction with the virtual world, at least with near real time control of the viewpoint is a critical test for a virtual reality. [Isdale, 1993]

Some people object to the term "virtual reality", saying it is an oxymoron. Other terms that have been used are synthetic environments, cyberspace, artificial reality, simulator technology, etc. Virtual reality, however, has caught most attention of the media (see, e. g., [Vince, 1995]).

The applications being developed for virtual reality run a wide spectrum, from games to architectural and business planning. Many applications are worlds that are very similar to our own, like CAD or architectural modeling. Some applications provide ways of viewing from an advantageous perspective not possible with the real world, like scientific simulators and telepresence systems, or air traffic control systems. Other applications are much different from anything we have ever directly experienced before. These latter applications (e. g. modeling the stock market, simulating large traffic control systems during rush hours) may be the systems that are most interesting but hardest to realize.

1.2 Historic Development of Virtual Reality Systems

A major distinction of virtual reality systems is the mode with which they interface to the user. The following list describes some of the common modes used in virtual reality systems.

- **Window on world systems (WoW)**: Some systems use a conventional computer monitor to display the visual world. This is sometimes called desktop virtual reality or a window on a world (WoW). This concept traces its lineage back through the entire history of computer graphics. In 1965, Ian Sutherland laid out a research program for computer graphics in a paper called "The Ultimate Display" that has driven the field for the past nearly thirty years.

  "One must look at a display screen," he said, "as a window through which one beholds a virtual world. The challenge to computer graphics is to make the picture in the window look real, sound real and the objects act real." [Sutherland, 1965]

- **Video mapping**: A variation of the WoW approach merges a video input of the user's silhouette with a 2D computer graphic. The user watches a monitor that shows his body's interaction with the world. [Kruger, 1983]

- **Immersive systems and CAVEs**: The ultimate virtual reality systems completely immerse the user's personal viewpoint inside the virtual world. These "immersive" virtual reality systems are often equipped with a head mounted display (HMD). This is a helmet or a face mask that holds the visual and auditory displays. The helmet may be free ranging, tethered, or it might be attached to some sort of a boom armature.

  A more sophisticated, but also more cost-expensive variation of the immersive systems uses multiple large projection displays to create a cave or room in which the viewer(s) stand. An early implementation was called "the closet cathedral" for the ability to create the impression of an immense environment within a small physical space [Isdale, 1993]. The *Computer Automated Virtual Environment* (*CAVE*, see [Holzer and Pirngruber, 1997]) is a commercial immersive system that has been installed several dozen times up to now [Sperlich and Schaermeli, 1997]. The standard CAVE offers a 3D stereoscopic projection onto three to five walls of a cube with a length of approximately 3 m. Thus, groups of visitors to the virtual world can enter the CAVE at the same time, led by one "guide" whose position is tracked. One or more graphic workstations generate – in real time – the images onto the virtual world according to the viewpoint of the tracked visitor. A sample CAVE layout is depicted in Fig. 1.1.

- **Telepresence**: Telepresence is a variation on visualizing complete computer generated worlds. This technology links remote sensors in the real world with the senses of a human operator. The remote sensors might be located on a robot, or similar. Fire fighters use remotely operated vehicles to handle some dangerous conditions. Surgeons are using very small instruments on cables to do surgery without cutting a major hole in their patients. The instruments have a small video camera at the business end. Robots equipped with telepresence systems have already changed the way deep sea and volcanic exploration is done. NASA used telerobotics for space exploration during the "sojourner" project for exploring the surface of the planet Mars, see [Wunsch, 1998].

- **Mixed reality**: Merging the telepresence and virtual reality systems gives mixed reality or seamless simulation systems. In this case the computer generated inputs are

Figure 1.1: A sample layout of a CAVE environment

merged with telepresence inputs and / or the users view of the real world. For instance, a surgeon's view of a brain surgery is overlaid with images from earlier MRT scans and real-time ultrasound. A fighter pilot sees computer generated maps and data displays inside his helmet visor or on cockpit displays.

- **Fish tank virtual reality**: The phrase "fish tank virtual reality" is used to describe a virtual reality system that combines a stereoscopic monitor display using LCD Shutter glasses with a mechanical head tracker [Ware *et al.*, 1993]. The resulting system is superior to simple stereo-WoW systems due to the motion parallax effects introduced by the head tracker. It thus combines the impression of a CAVE with the cost-effectiveness of a WoW system.

1.3 Virtual Reality Hardware Components

1.3.1 Hardware Devices

There exist a number of specialized types of hardware devices that have been developed or used for virtual reality applications. The most important are listed below.

Image Generator

One of the most time consuming tasks in a virtual reality system is the generation of the images. Fast computer graphics opens a very large range of applications aside from virtual reality, so there has been a market demand for hardware acceleration for a long while. There are currently a number of vendors selling image generator cards for PC level machines, many of these are based on the OpenGL standard [Neider *et al.*, 1993] developed by SGI. SGI Inc. (formerly Silicon Graphics Inc.) has made a very profitable business of producing graphics workstations. SGI workstations are some of the most common computers found in virtual reality laboratories.

Manipulation and Control Device

One key element for interaction with a virtual world is a means of tracking the position of a real world object, such as a head or hand. There are numerous methods for position tracking and control. Ideally a technology should provide 3 measures for position (X, Y, Z) and 3 measures of orientation (roll, pitch, yaw). One of the biggest problem for position tracking is latency, or the time required to make the measurements and preprocess them before input to the simulation engine.

The simplest control hardware is a conventional mouse, trackball or joystick. While these are two-dimensional devices, creative programming can use them for 6D controls. There are, additionally, a number of 3 and 6 dimensional mice / trackball / joystick devices available at the market. These add some extra buttons and wheels that are used to control not just the XY translation of a cursor, but its Z dimension and rotations in all three directions.

One common virtual reality device is the instrumented glove. Such a glove is outfitted with sensors on the fingers as well as an overall position / orientation tracker, see Fig. 1.2 [Willim, 1989]. There are a number of different types of sensors that can be used. VPL made several DataGloves, mostly using fiber optic sensors for finger bends and magnetic trackers for overall position. Mattel manufactured the PowerGlove for use with the Nintendo game system, for a short time, see Fig. 1.3 [Bernatchez, 1995]. This device is easily adapted to interface to a personal computer. It provides some limited hand location and finger position data using strain gauges for finger bends and ultrasonic position sensors.

The concept of an instrumented glove has been extended to other body parts. Full body suits with position and bend sensors have been used for capturing motion for character animation, control of music synthesizers, etc. in addition to virtual reality applications. [Craiger, 1999]

Position Tracking

Mechanical armatures can be used to provide fast and very accurate tracking. Such armatures may look like a desk lamp (for basic position / orientation) or they may be highly complex exoskeletons (for more detailed positions). The drawbacks of mechanical sensors are the encumbrance of the device and its restrictions on motion.

Magnetic trackers use sets of coils that are pulsed to produce magnetic fields. The magnetic sensors determine the strength and angles of the fields. Limitations of these trackers are a high latency for the measurement and processing, range limitations, and interference from ferrous materials within the fields.

Ultrasonic sensors can be used to track position and orientation. A set of emitters and receivers are used with a known relationship between the emitters and the receivers. The emitters are pulsed in sequence and the time lag to each receiver is measured. Triangulation gives the position. Drawbacks to ultrasonics are low resolution, long lag times and interference from echoes and other noises in the environment. Nevertheless, most of the modern tracking systems are based upon ultrasonic sensors like, e. g., the InterSense 6D tracking system [InterSense, 1999].

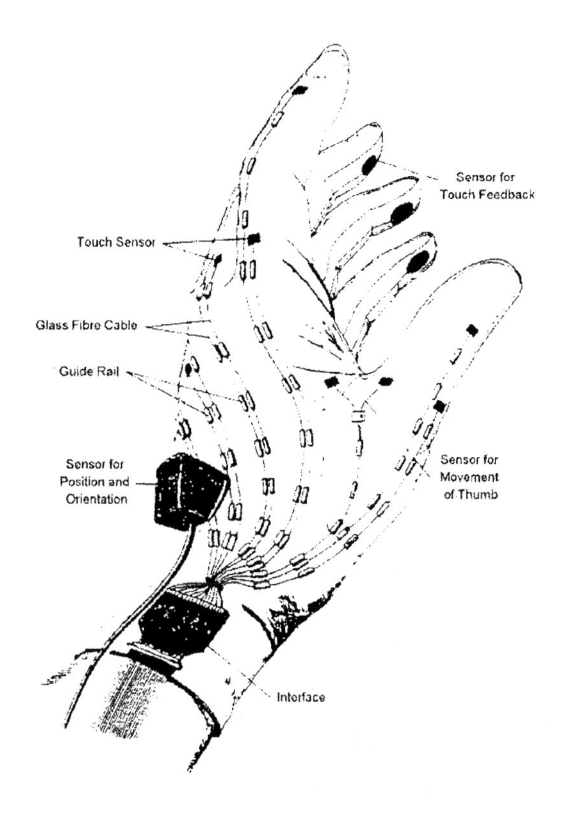

Figure 1.2: Principal design of a data glove

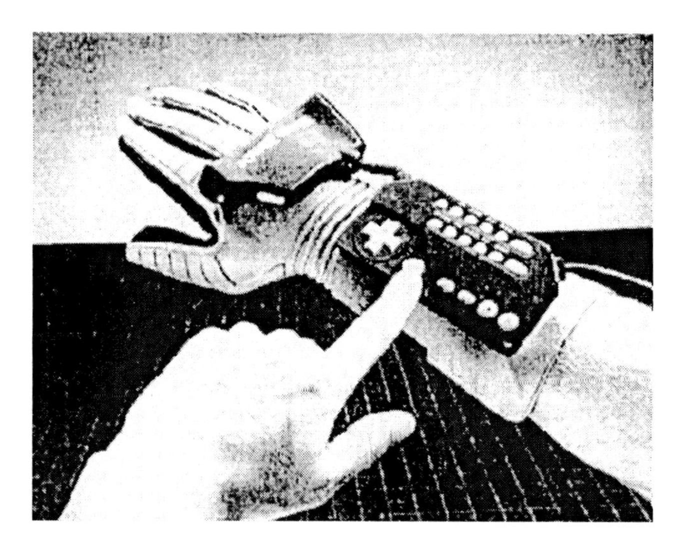

Figure 1.3: The PowerGlove

Stereo Vision

Stereo vision is often included in a virtual reality system. This is accomplished by creating two different images of the world, one for each eye. The images are computed with the viewpoints offset by the equivalent distance between the eyes. There are a large number of technologies for presenting these two images. The images can be placed side-by-side and the viewer asked (or assisted) to cross their eyes. The images can be projected through differently polarized filters, with corresponding filters placed in front of the eyes. Anaglyph images use red / blue glasses to provide a monochrome stereo vision.

Alternatively, the two images can be displayed sequentially on a conventional monitor or projection display. Liquid Crystal shutter glasses are then used to shut off alternate eyes in synchronization with the display. When the brain receives the images in rapid enough succession, it fuses the images into a single scene and perceives depth. A fairly high display swapping rate (min. 60 Hz) is required to avoid perceived flicker.

Another alternative method for creating stereo imagery on a computer is to use one of several split screen methods. These divide the monitor into two parts and display left and right images at the same time. One method places the images side by side and conventionally oriented. It may not use the full screen or may otherwise alter the normal display aspect ratio. A special hood viewer is placed against the monitor which helps to position the eyes correctly and may contain a divider so each eye sees only its own image. [Isdale, 1993]

Head Mounted Display

One hardware device closely associated with virtual reality is the head mounted display (HMD), see Fig. 1.4 [Bernatchez, 1995]. Such a device uses some sort of helmet or goggles to place small video displays in front of each eye, with special optics to focus and stretch the perceived field of view. Most HMDs use two displays and can provide stereoscopic imaging. A HMD requires a position tracker in addition to the helmet. Alternatively, the display can be mounted on an armature for support and tracking (a boom display).

1.3.2 Levels of Virtual Reality Hardware Systems

The following list defines a number of levels of virtual reality hardware systems. These are not hard levels, especially towards the more advanced systems.

- **Entry virtual reality system**: The entry level virtual reality system takes a stock personal computer or workstation and implements a WoW system. It includes a graphic display, a 2D input device like a mouse, trackball or joystick, the keyboard, hard disk and memory.
- **Basic virtual reality system**: The next step up adds some basic interaction and display enhancements. Such enhancements would include a stereographic viewer (LCD shutter glasses) and a input / control device such as a data glove and / or a multidimensional (3D or 6D) mouse or joystick.

Figure 1.4: Example of a head mounted display helmet

- **Advanced virtual reality system**: The next step is to add a rendering accelerator and / or frame buffer and possibly other parallel processors for input handling, etc. The simplest enhancement in this area is a faster display card. For the PC class machines, there are a number of fast graphic accelerator cards, mostly based upon OpenGL hardware support. These can make a dramatic improvement in the rendering performance of a desktop virtual reality system.

  An advanced virtual reality system might also add a sound card to provide mono, stereo or true 3D audio output. Some sound cards also provide voice recognition. This constitutes an excellent additional input device for virtual reality applications.

- **Immersive virtual reality system**: An immersive virtual reality system adds some type of immersive display system: a HMD, a boom device, or multiple large projection type displays (CAVE).

  An immersive virtual reality system might also add some form of tactile, haptic and touch feedback interaction mechanisms. The area of touch or force feedback (known collectively as haptics) is a rather new research area, see, e. g., [Srinivasan and Basdogan, 1997].

- **Cockpit simulator**: A common variation on virtual reality is to use a cockpit or cab compartment to enclose the user. The virtual world is viewed through some sort of view screen and is usually either projected imagery or a conventional monitor. The cockpit simulation is very well known in aircraft simulators, with a history dating back to the early 1930's. The cockpit is often mounted on a motion platform that can give the illusion of a much larger range of motion.

1.4 Virtual Reality Software Components

Just what is required of a virtual reality program? The basic parts of the system (see Fig. 1.5) can be broken down into

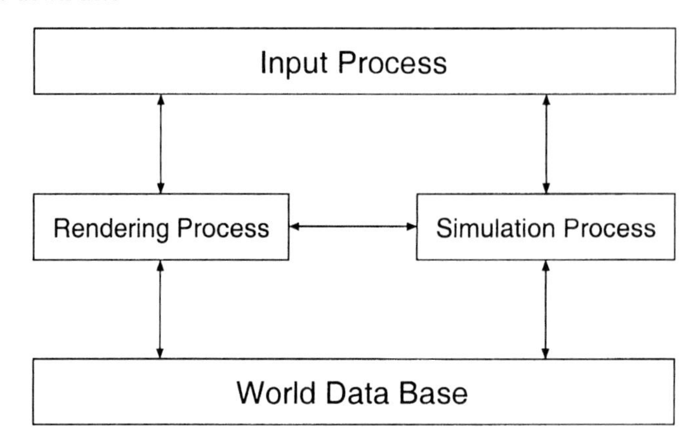

Figure 1.5: Basic software components of a virtual reality system

- input processes,
- a simulation processor,
- rendering processes, and
- a virtual world description.

All these parts must consider the time required for processing. Every delay in response time degrades the feeling of "presence" and reality of the simulation.

1.4.1 Input Processes

The input processes of a virtual reality program control the devices used to input information to the computer. There exist a wide variety of possible input devices: keyboard, mouse, trackball, joystick, 3D and 6D position trackers (glove, wand, head tracker, body suit, etc.). A networked virtual reality system adds input received from net. A voice recognition system is also a good augmentation for virtual reality, especially if the user's hands are being used for other tasks. Generally, the input processing of a virtual reality system is kept simple. The object is to get the coordinate data to the rest of the system with minimal lag time. Some position sensor systems add some filtering and data smoothing processing. Some glove systems add gesture recognition. This processing step examines the glove inputs and determines when a specific gesture has been made. Thus it can provide a higher level of input to the simulation.

1.4.2 Simulation Processor

The core of a virtual reality program is the simulation system. This is the process that knows about the objects and the various inputs. It handles the interactions, the scripted object actions, simulations of physical laws (real or imaginary) and determines the world status. This simulation is basically a discrete process that is iterated once for each time step or frame. A networked virtual reality application may have multiple simulations running on different machines, each with a different time step. Coordination of these can be a complex task.

The simulation engine takes the user inputs along with any tasks programmed into the world such as collision detection, scripts, etc. and determines the actions that will take place in the virtual world.

1.4.3 Rendering Processes

The rendering processes of a virtual reality program are those that create the sensations that are output to the user and to other network processes. There exist separate rendering processes for visual, auditory, haptic (touch / force), and other sensory systems. Each renderer takes a description of the world state from the simulation process or derives it directly from the world database for each time step.

- **Visual rendering**: The visual renderer is the most common process and it has a long history from the world of computer graphics and animation. The major consideration of a graphic renderer for virtual reality applications is the frame generation rate. It is necessary to create a new frame every 1/20 of a second or faster. 20 frames per second is roughly the minimum rate at which the human brain will merge a stream of still images and perceive a smooth animation.

  An effective short cut for visual rendering is the use of textures or image maps. These are pictures that are mapped onto objects in the virtual world. Instead of calculating lighting and shading for the object, the renderer determines which part of the texture map is visible at each visible point of the object. The resulting image appears to have significantly more detail than is otherwise possible.

- **Auditory rendering**: A virtual reality system is greatly enhanced by the inclusion of an audio component. This may produce mono, stereo or 3D audio. The latter is a fairly difficult proposition. It is not enough to do stereo-pan effects as the mind tends to locate these sounds inside the head. Research into 3D audio has shown that there are many aspects of our head and ear shape that effect the recognition of 3D sounds. It is possible to apply a rather complex mathematical function (called a head related transfer function or HRTF) to a sound to produce this effect, see [Isdale, 1993]. The HRTF is a very personal function that depends on the individual's ear shape, etc. However, there has been significant success in creating generalized HRTFs that work for most people and most audio placement. There remains a number of problems, such as the "cone of confusion" wherein sounds behind the head are perceived to be in front of the head.

  Sound has also been suggested as a means to convey other information, such as surface roughness. Dragging your virtual hand over sand would sound different than dragging it through gravel.

- **Haptic rendering**: Haptics is the generation of touch and force feedback information. This area is a very new science and there is much to be learned. There have been very few studies done on the rendering of true touch sense (such as liquid, fur, etc.). Almost all systems to date have focused on force feedback and kinesthetic senses. These systems can provide good clues to the body regarding the touch sense, but are considered distinct from it. Many of the haptic systems thus far have been exo-skeletons that can be used for position sensing as well as providing resistance to movement or active force application.

- **Other senses**: The sense of balance and motion can be served to a fair degree in a virtual reality system by a motion platform. These are used in flight simulators and theaters to provide some motion cues that the mind integrates with other cues to perceive motion. It is not necessary to recreate the entire motion perfectly to fool the mind into a willing suspension of disbelief.

  The sense of temperature has seen some technology developments. There exist very small electrical heat pumps that can produce the sensation of heat and cold in a localized area. These system are fairly expensive.

  Other senses such as taste, smell, pheromone, etc. are beyond the current ability to render rapidly and effectively. Sometimes, we just do not know enough about the functioning of these other senses.

1.4.4 Virtual World Description

The virtual world itself needs to be defined in a "world space". By its nature as a computer simulation, this world is necessarily limited. The computer must put a numeric value on the locations of each point of each object within the world. Usually these coordinates are expressed in Cartesian dimensions of X, Y, and Z (length, height, depth).

One method of dealing with the limitations on the world coordinate space is to subdivide a virtual world into multiple worlds and provide a means of transiting between the worlds. This allows fewer objects to be computed both for scripts and for rendering and necessitates multiple stages (a. k. a. rooms, areas, zones, worlds, multiverses, etc.) and a way to move between them (so-called portals).

1.5 Virtual World Modeling

1.5.1 World Database

The storage of information on objects and the world is a major part of the design of a virtual reality system. The primary things that are stored in the world database (or world description files) are the objects that inhabit the world, scripts that describe actions of those objects or the user (things that happen to the user), lighting, program controls, and hardware device support.

1.5.2 Object Behavior

A virtual world consisting of solely static objects is only of mild interest. Many researchers and enthusiasts of virtual reality have remarked that interaction is the key to a successful and interesting virtual world. This requires some means of defining the actions that objects take on their own and when the user (or other objects) interact with them. This concept is generally referred to as world scripting, subdivided into three basic types:

1. **Motion scripts** modify the position, orientation or other attributes of an object, light or camera based on the current system tick. A *tick* is one advancement of the simulation clock. Generally, this is equivalent to a single frame of visual animation. (Virtual reality generally uses discrete simulation methods.)

   For simplicity and speed, only one motion script should be active for an object at any one instant. Motion scripting is a potentially powerful feature, depending on how complex these scripts may become. Care must be exercised since the interpretation of these scripts will require time, which impacts the frame and delay rates.

   Additionally, a script might be used to attach or detach an object from a hierarchy. For example, a script might attach the user to a virtual car when he wishes to drive around the virtual world. Alternatively, the user might pick up or attach an object to himself.

2. **Trigger scripts** are invoked when some trigger event occurs, such as collision, proximity or selection. The virtual reality system needs to evaluate the trigger parameters at each tick. For proximity detectors, this may be a simple distance check from the object to the 3D eye or virtual human (a. k. a. *avatar*). Collision detection is a process that needs more computational effort, see Ch. 12.3.

3. **Connection scripts (mediators)** control the connection of input and output devices to various objects. For example a connection script may be used to connect a glove device to a virtual hand object. The glove movements and position information are used to control the position and actions of the hand object in the virtual world. Some systems build this function directly into the program. Other systems are designed such that the virtual reality program is almost entirely a connection script.

Examples of scripting languages are VRML-Script, JavaScript and pure Java in combination with an embedded Java Virtual Machine, see [Flanagan, 1998].

1.5.3 User Interface System

A virtual reality system often needs to have some sort of control panels available to the user. The world database may contain information on these panels and how they are integrated into the application. Alternatively, they may be a part of the program code. There are several ways to create these panels. There could be 2D menus that surround a WoW display, or are overlaid onto the image. An alternative is to place control devices inside the virtual world. The simulation system must then note user interaction with these devices as providing control over the world.

One primary area of user control is control of the viewpoint (moving around within the virtual world). Some systems use the joystick or similar device to move. Others use gestures from a glove, such as pointing, to indicate a motion command.

The user interface to the virtual world might be restricted to direct interaction in the 3D world. However, this is extremely limiting and requires lots of 3D calculations. Thus it is desirable to have some form of 2D Graphic user interface to assist in controlling the virtual world. These control panels of the would appear to occlude portions of the 3D world, or perhaps the 3D world would appear as a window or viewport set in a 2D screen interface. The 2D interactions could also be represented as a flat panel floating in 3D space, with a 3D effector controlling them.

1.5.4 Feedback on User Interaction

The user must be given some indication of interaction feedback when the virtual cursor selects or touches an object. Crude systems have only the visual feedback of seeing the cursor (virtual hand) penetrate an object. The user can then grasp or otherwise select the object. The selected object is then highlighted in some manner. Alternatively, an audio signal could be generated to indicate a collision. Some systems use simple touch feedback, such as a vibration in the joystick, to indicate collision, etc.

1.5.5 VRML – The Virtual Reality Modeling Language

The Virtual Reality Modeling Language (VRML) has been developed out of the need to include 3D interfaces into the World Wide Web (WWW) [Hase, 1997]. The goal was to develop a language for describing objects in 3-space, visualize these objects using standard browsers on a computer screen and interconnect the such described virtual world with the Internet.

The concept of VRML was born during the first international conference concerning the WWW, held at CERN in Geneva, Switzerland, in May 1994, where a session was organized under the topic "A virtual reality markup language and the World Wide Web" [Jenewein, 1999]. The session was coached by Tim Berners-Lee, the inventor of HTML. VRML was then developed installing Internet discussion forums and checking available languages for the modeling of virtual worlds. Soon OpenInventor from SGI, an object-oriented graphics library based upon OpenGL whose core was put into the public domain, was chosen as a basis and extended into VRML 1.0, which was presented in October 1994. However, VRML 1.0 completely lacked the concept of a scripting language for modeling dynamic behavior, among other deficits.

Consequently, in August 1996 VRML 2.0 [Laura *et al.*, 1996] was presented (again based loosely upon an SGI product called "Moving Worlds"). VRML 2.0 includes interfaces to JavaScript and Java in addition to supporting its own scripting language, VRMLScript. It also allows textures based upon commonly used picture formats and supports an include mechanism that allows to break down large virtual worlds into smaller components, which improves reuse and optimization of virtual worlds considerably. VRML 2.0 was standardized by ISO under the term "VRML 97" [Hase, 1997].

Currently there is some consideration, whether VRML will continue to be the main modeling language for virtual worlds (VRML 2000 project) or will be replaced by either Java3D or X3D, an extension of XML.

1.6 Authoring Systems

A virtual world can be created, modified and experienced. Some virtual reality systems may not distinguish the creation and experiencing aspects. However, there is currently a much larger body of experience to draw upon for designing the world from the outside. This method may use techniques borrowed from architectural and other forms of computer aided design (CAD) systems. Also the current technologies for immersive virtual reality systems are fairly limiting in resolution, shading, lighting models, etc. They are not nearly as well developed as those for more conventional computer graphics and interfaces.

For many virtual reality systems, it makes a great deal of sense to have an *authoring mode* and a *playback mode*. The authoring mode may be a standard text editor and compiler system, or it may include 3D graphic and other tools. Such a split mode system makes it easier to create a stand alone application that can be delivered as a product.

An immersive authoring ability may be desirable for some applications and some users. For example, an architect might have the ability to move walls, etc. when immersed, while the clients with him, who are less familiar with the system, are limited to player status. That way they can't accidentally rearrange the house by leaning on a wall.

Chapter 2

Computer Aided Design and Modeling

This chapter describes the origins of graphic-data processing and its evolution into computer aided design (CAD) and computer aided manufacturing (CAM). Additionally, the coupling of CAD and CAM into computer supported production is discussed.

2.1 Graphic-Data Processing

There exist various fields in data processing, where the treatment of images and image information is of interest. The following areas are the main branches processing graphic information (Fig. 2.1):

- **Generative graphic-data processing**: Images are created, manipulated and drawn taking internal descriptions.
- **Image processing**: "Bad-quality" images are improved to make a further processing possible, e. g. by a human or a computer (such improvements are, for instance, error elimination, contrasting or filtering).
- **Pattern recognition**: Certain information is gained from an image leading to an internal description of the image.

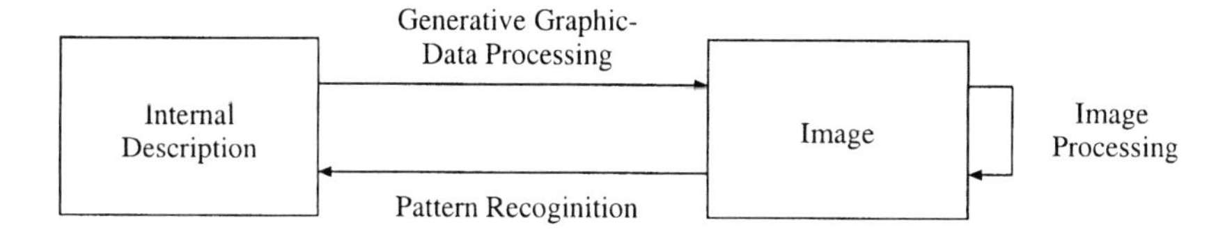

Figure 2.1: Graphic-data processing

In the following, we restrict the term "graphic-data processing" to generative graphic-data processing, where we include the fact of interaction as a main facility.

The importance of graphic-data processing can be described best by listing some applications. We know that the following list is – by far – not complete, but it shall give an impression of the wide fields where graphic-data processing is used:

- computer aided design,
- medical applications (cardiograms, tomographies, etc.),
- cartography (often military applications),
- business graphics,
- simulation (e. g. flight simulators),
- computer games,
- animation (trick films),
- computer art.

2.2 Scope of Computer Aided Design

The meaning of computer aided design (CAD) has changed several times in its past twenty years of history. For some time, CAD was almost synonymous with finite element structural analysis. Later, the emphasis shifted to computer aided drafting (most commercially available CAD systems are actually drafting systems). More recently, CAD has been associated with the design of three-dimensional objects (this is typical in many branches of mechanical engineering). We consider *CAD* as a discipline that provides the required know-how in computer hardware and software, in systems analysis and in engineering methodology for specifying, designing, implementing, introducing and using computer based systems for design purposes.

Computer aided design is often treated together with computer aided manufacturing (CAM). CAM starts from data – preferable machine-readable data – that are produced in the design process, but CAM is not part of the design process itself. The same applies to computer aided testing (CAT) and computer aided process planning (CAP).

Recently, the term computer aided engineering (CAE) has been used for summarizing all computer aids in design, while restricting CAD to computer aided drafting, Here, however, we will continue to associate the term CAD with the wider meaning defined above.

Design is not only the more-or-less intuitively guided creation of new information by the designer. It also comprises analysis, presentation of results, simulation and optimization. These are essential constituents of the iterative process, leading to a feasible and, one hopes, optimal design.

There are several reasons for implementing a CAD system. Nevertheless, the following arguments seem to be fundamental:

1. **Increasing the productivity of the designer**: This is accomplished by helping the designer to visualize the product and its component sub-assemblies and parts; and

by reducing the time required in synthesizing, analyzing, and documenting the design. This productivity improvement translates not only into lower design costs but also into shorter project completion times.

2. **Improving the quality of the design**: A CAD system permits a more thorough engineering analysis and a larger number of design alternatives can be investigated. Design errors are also reduced through the greater accuracy provided by the system. These factors lead to a better design.

3. **Improving communication**: Use of a CAD system provides better engineering drawings, more standardization in the drawings, better documentation of the design, fewer drawing errors, and greater legibility.

4. **Creating a data base for manufacturing**: During the process of creating the documentation for the product design (geometries and dimensions of the product and its components, material specifications for components, bill of materials, etc.), much of the required data base to manufacture the product is also created.

2.3 History of CAD

In this section, we will give a brief review of the historical background of CAD. Knowledge about the history provides a better understanding of the present state of the art, and may even enhance the creativity of those planning to work in this field.

1. Early in the 1950s, the Servomechanisms Laboratory at the Massachusetts Institute of Technology (M.I.T.) developed the first automatically controlled milling machine using the Whirlwind computer. We note that computer aided manufacturing is not a descendant of CAD, but has a distinct origin of its own.

2. The activities of M.I.T. led to the evolution of the Automatically Programmed Tool (APT). The step from APT to design programs including computer graphics functions was outlined by Coons.

3. Ian Sutherland, one of the first CAD pioneers, envisaged the designer sitting in front of a console using interactive graphics facilities developed at the M.I.T.; he developed SKETCHPAD in the beginning of the 1960s [Sutherland, 1965]. The software principles of rubber band lines, circles of influence, magnification, rotation, and sub-framing were born in those days.

4. In 1964, General Motors announced the DAC-1 (Design Augmented by Computer) system. DAC-1 was more concerned with producing hard-copies of drawings than with interactive graphic techniques.

5. In 1965, Bell Telephone Laboratories announced the GRAPHIC 1 remote display system. The system was used for geometrically arranging printed-circuit components and wirings, for the composing and editing of text, and for the interactive placement of connective wiring. It was a very early implementation of the important idea of having the CAD processing power distributed among local interactive workstations and a central host computer.

6. Freeman suggested, in 1967, an algorithm for the solution of hidden-line problems.

7. A system called GOLD was developed in 1972 at RCA for integrated circuit mask layout.

8. The first half of the Seventies was a time of much enthusiasm among the early CAD scientists and system developers. Much theoretical work was done, laying down the fundamentals of CAD as we know it today. The Integrated Civil Engineering System (ICES) was developed, followed by a number of systems which implemented many principal ideas regarding a CAD methods base. The theory of finite elements and associated programs started a booming development. At the same time, considerable research activity was going on in the area of hidden line / surface removal.

9. The University of Rochester started the Production Automation Project in 1972. As a result of this project, two geometric modeling systems, PADL-1 and PADL-2, were developed. These system enables to model objects as "solids", i. e. a model that is complete and unambiguous, and it is assured that only physically realizable parts can be modeled.

10. In 1973, a Lockheed review demonstrated that computer graphics will not only be practicable in the design process, but also cost effective.

11. Hewlett-Packard announced, in 1978, a microprocessor-based raster scan display terminal.

12. Several publications by General Motors and Boeing in 1978 confirmed the usefulness of CAD / CAM technology and described how to bridge the gap between CAD and CAM. The late Seventies may be characterized as the time of CAD's breakthrough from a scientific endeavor to an economically attractive and – in many areas – indispensable tool in industry.

13. The Eighties saw CAD fully developed in the market place, and it became a standard tool in all design offices, progressing in tandem with a steady adaptation of the work procedures there.

14. In the Nineties, further development was comprised of "full 3D" CAD, fault-tolerant CAD, "intelligent" CAD that incorporates techniques of knowledge engineering, and integrating CAD into an enterprise-wide computer support.

2.4 Structure of CAD Systems

2.4.1 Software Components

The graphic-interactive communication capabilities of a CAD system are supported by tools like geometry processing, graphics processing, etc. The results are stored in the computer internal model. With these characteristics CAD software can be structured as illustrated in Fig. 2.2.

This logical structure divides CAD-software into four parts which communicate with each other:

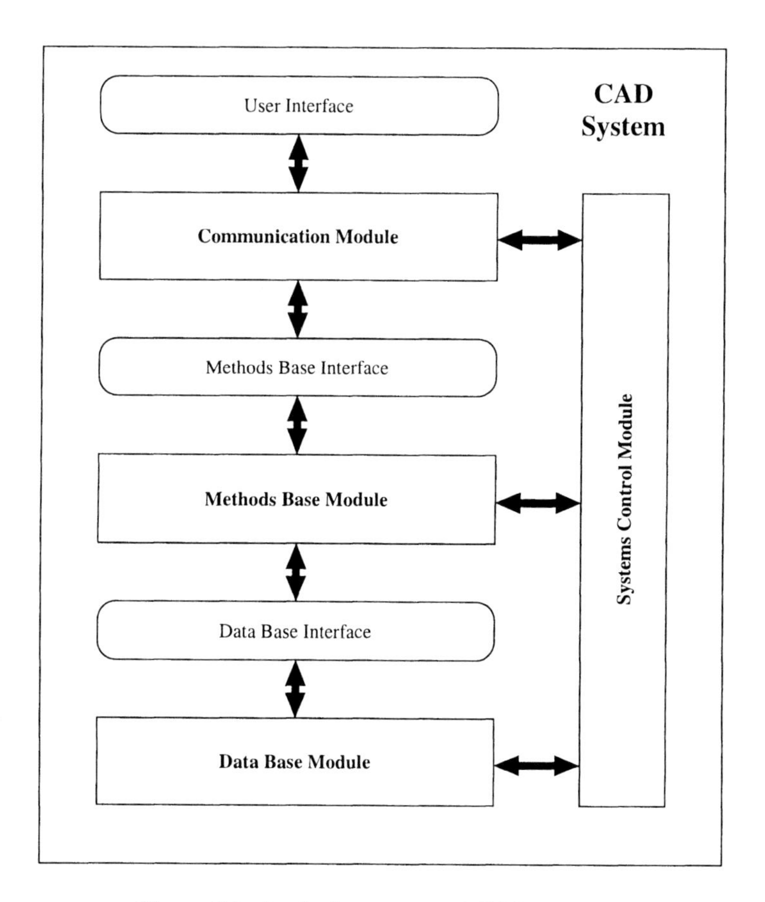

Figure 2.2: Logical structure of CAD systems

1. The *communication module* is the interface between designer and CAD system. It contains commands, controls, graphic-interactive input / output functions and initialization methods with assigned data.

2. The *methods base module* provides a number of specific methods to solve design tasks. Methods contain algorithms including instructions about the performance of the algorithms, their scope and applications. Typical methods are modules for geometry processing, calculation, dimensioning, tolerancing, etc.

3. The *data base module* provides a functional interface to the data base manipulation language to manipulate (insert, delete, modify, and get) product definition data. Product definition data are stored as a computer internal model.

4. The *control module* controls and coordinates the different modes of the system.

The logical structure allows definition of separate modules, which perform specific tasks as needed. The purpose of this structure is to make CAD software more flexible and efficient. These modules may also be realized with hardware.

2.4.2 The Computer Internal Model

The technical solution of a product is represented by the *computer internal model*. The purpose of the computer internal model is to provide product definition data for the different design functions such as:

- modifying, optimizing, finishing, and testing of the technical solution of the designed product,
- generating production schedules and part lists,
- generating programs for NC-machines and robots,
- simulating manufacturing and assembly operations.

Computer internal models are usually stored in and controlled by a data base management system. When the product definition data are stored separately from processing data the flexibility of the software is improved. Thus, data for processing methods can be added and updated. Changes to these processing methods alter the computer internal model and make modifications or additions necessary. With the concept of a CAD system which is based on the computer internal model further applications can be integrated. The degree of integration or addition of other production planning and control activities depends on the complexity of the computer internal model used in the CAD system.

Different Geometric Models for CAD

The geometric representation is one of the main components of the computer internal model. Computer internal models can be structured as a matrix where the types of generated model data (digital cell models, topological models) are listed in rows and the computer internal representation (procedure internal, data structure) in columns. The geometry of a product can be represented in steps of different complexity reaching from two-dimensional line geometry up to a solid representation of the product. Another classification divides the geometric elements into analytical and numerical ones. Fig. 2.3 shows geometric model elements of different complexity.

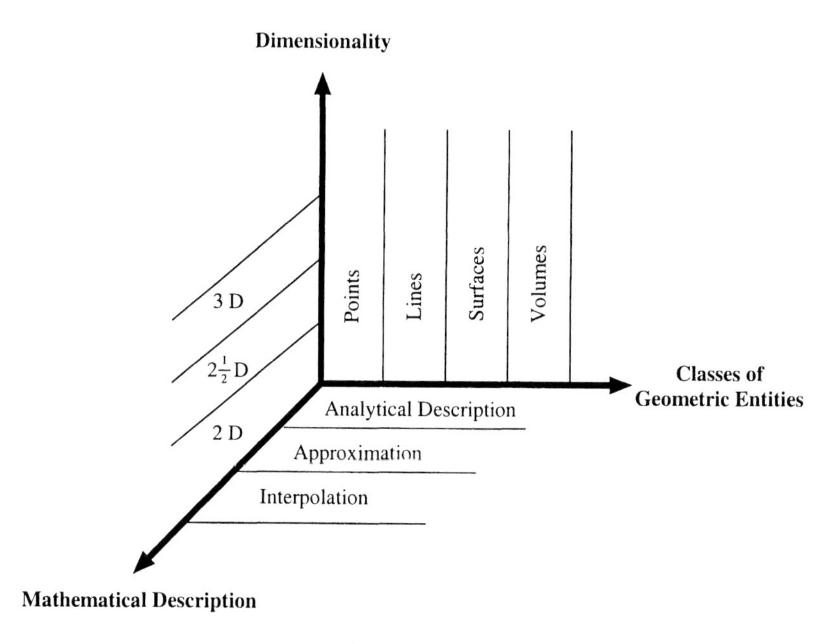

Figure 2.3: Geometric model elements of different complexity

A feature of the computer internal model is its parametric capability which allows the change of dimensions for a defined geometry whereby the topology of the product is maintained. With parametric models product variants can be generated easily. To use parametric models efficiently, algorithms for automatic dimensioning and tolerancing of the referenced geometric element are required.

In addition to the geometric model, other important information is generated during the design process. One essential output is the technological description of the product. When the modeling process is considered as an entity a computer internal product model may be structured into four "sub-models" which are related to each other.

1. The *user model* contains elements of the user application and the history of the design process.

2. The *geometry model* contains the product geometry represented by a volume model (constructive solid modeling, boundary representation or something similar).

3. The *drawing model* is generated from the geometry model and is the two-dimensional representation of the product.

4. The *technology model* provides technological information and supplements the geometric representation needed for the generation of manufacturing and assembly data.

It is important that in addition to the model data the semantics of the data and the relations between the data are stored.

2.4.3 Integration of the Production Planning Process

When production planning and control functions are automated they should be integrated into the CAD system to utilize existing product definition data. The connection of these functions should be done via a defined interface, which means that product definition data have to be transformed to a specified format. Any additional application should be able to process the format of that interface. the integration is based on the common product definition data, for example as a computer internal model stored in a data base. Communication techniques and processing methods for the special process are integrated into the system and access data from the common definition (Fig. 2.4).

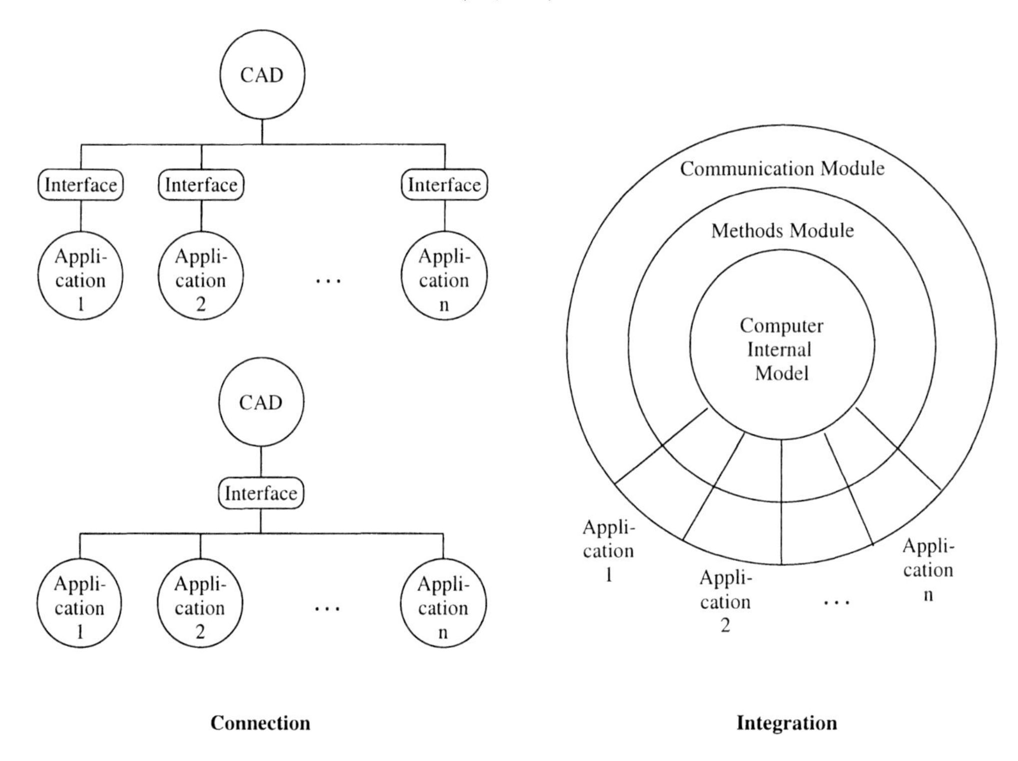

Figure 2.4: Connection versus integration of production planning and control functions to a CAD system

The connection and integration of further production planning and control functions to a CAD system can be achieved by the following different tasks:

1. preparation of design data for the manufacturing process (for example NC-oriented dimensioning of a part),
2. generation of standard interfaces (for example STEP / PDES),

3. generation of the NC-program (for example the part description in an NC-language format),
4. generation of complete production information.

These tasks allow different degrees of automation to support manufacturing planning and NC-programming. Additional work may have to be done if the support does not satisfy all manufacturing requirements.

The integration of production planning and control functions will result in the following advantages:

1. exchange of complete production definition data by digital means,
2. computer aided preparation of technical data,
3. reduction of manual effort,
4. avoidance of duplicated work,
5. computer support for the control of technical data,
6. reduction of test and control procedures,
7. improved engineering productivity,
8. shorter lead times,
9. reduced engineering personnel requirements,
10. easier customer modifications,
11. faster response to requests for quotations,
12. avoidance of sub-contracting to meet schedules,
13. minimized transcription errors,
14. improved accuracy of design,
15. easier recognition of component interactions in analysis,
16. better functional analysis to reduce prototype testing,
17. assistance in preparation for documentation,
18. more standardization of designs ,
19. better designs provided,
20. improved productivity in tool design,
21. better knowledge of costs provided,
22. reduced training time for routine drafting tasks and NC part programming,
23. fewer errors in NC part programming,

24. potential for using more existing parts and tooling,
25. ensured appropriate design to existing manufacturing techniques,
26. materials and machining time saves by optimization algorithms,
27. availability of operational results on the status of work in progress,
28. increase in effectivity of the management of design personnel on projects,
29. assistance in inspection of complicated parts,
30. better communication interfaces and greater understanding among engineers, designers, drafters, management, and different project groups.

2.4.4 The Planning Process Based on CAD Models

The generation of production schedules is one of the main tasks of the manufacturing planning process. The process of generating the production schedule is based on the technical solution of the design process, the order requirements, the manufacturing know how and the financial means of the company. This planning process includes the selection of raw materials and the manufacturing sequence, the definition of machine tools and the generation of manufacturing times. A well designed CAD system should offer the following capabilities:

1. to make available the product definition data, including
   - geometric data,
   - technological data, and
   - organizational data

   in addition to the semantics and the relations of the workpiece attributes,
2. to provide a simulation capability to graphically model the manufacturing process including the tool path, machine tool workspace and fixture geometry,
3. to offer technological and organizational information,
4. to provide communication and efficient planning techniques.

With these capabilities the production schedule can be generated by graphic interactive communication techniques. The simulation of the planning process is done with the help of the computer internal model. The use of connection modules or common interfaces is another possibility to connect a CAD system with a CAP (computer aided planning) system.

2.5 The Design Process

2.5.1 Design – an Iterative Process

Before examining several facets of computer aided design, let us first consider the general design process. The process of designing something is characterized as an *iterative process* [Shigley, 1996], which consists of six identifiable steps or phases:

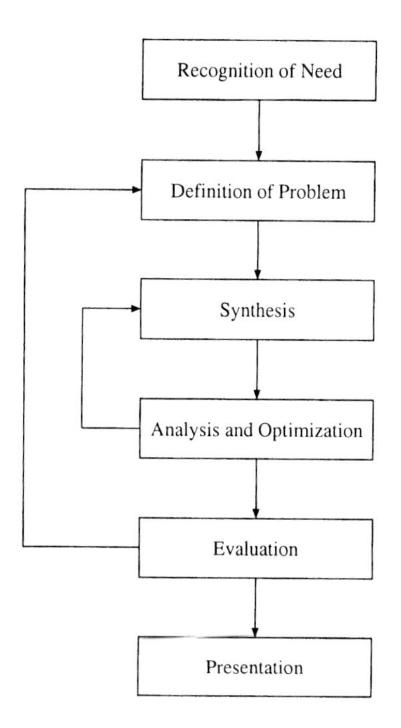

Figure 2.5: The general design process

1. recognition of need,
2. definition of the problem,
3. synthesis,
4. analysis and optimization,
5. evaluation,
6. presentation.

Recognition of need involves the realization by someone that a problem exists for which some corrective action should be taken. This might be the identification of some defect in a current machine design by an engineer or the perception of a new product marketing opportunity by a salesperson. *Definition of the problem* involves a through specification of the item to be designed. This specification includes physical and functional characteristics, cost, quality, and operating performance.

Synthesis and *analysis* are closely related and highly iterative in the design process. A certain component or subsystem of the overall system is conceptualized by the designer, subjected to analysis, improved through this analysis procedure, and redesigned. The process is repeated until the design has been optimized within the constraints imposed on the designer. The components and subsystems are synthesized into the final overall system in a similar iterative manner.

Evaluation is concerned with measuring the design against the specifications established in the problem definition phase. This evaluation often requires the fabrication and testing of a prototype model to assess operating performance, quality, reliability, and other criteria. The final phase in the design process is the *presentation* of the design. This step includes documentation of the design by means of drawings, material specifications, assembly lists, and so on. Essentially, the documentation requires a design data base to be created. Fig. 2.5 illustrates the basic steps in the design process, indicating its iterative nature.

2.5.2 Computer Aided Design

The various design-related tasks which are performed by a modern computer aided design system can be grouped into four functional areas:

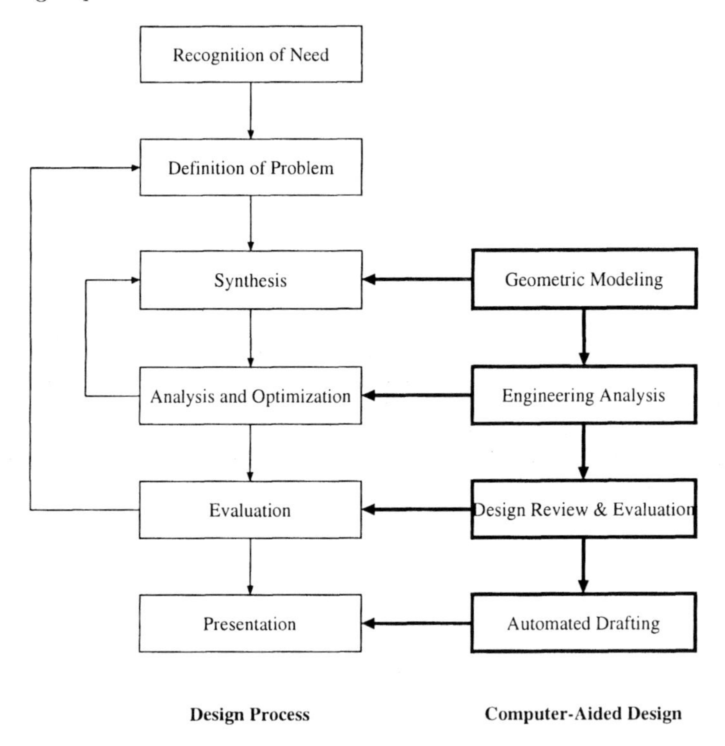

Figure 2.6: Application of computers to the design process

1. geometric modeling,
2. engineering analysis,
3. design review and evaluation,
4. automated drafting.

These four areas correspond to the final four phases in Shigley's general design process, illustrated in Fig. 2.6 [Groover and Zimmers, 1984]. *Geometric modeling* corresponds to the synthesis phase in which the physical design project takes form. *Engineering analysis* corresponds to phase 4, dealing with analysis and optimization. *Design review and evaluation* is the fifth step in the general design procedure. *Automated drafting* involves a procedure for converting the design image data residing in computer memory into a hard-copy document. It represents an important method for the presentation phase of the design.

2.6 The Product Cycle

2.6.1 Computer Aided Manufacturing

In order to analyze the scope of computer aided design and computer aided manufacturing (CAM) in the operation of a manufacturing firm, it is appropriate to examine the various activities and functions that must be accomplished in the design and manufacturing of a product. We will refer to these activities and functions as the *product cycle*.

A diagram showing the various steps in the product cycle is presented in Fig. 2.7 [Groover and Zimmers, 1984]. The cycle is driven by customers and markets that demand the product. It is realistic to think of these as a large collection of diverse industrial and consumer markets rather than one monolithic market. Depending on the particular customer group, there will be differences in the way the product cycle is activated. In some cases, the design functions are performed by the customer and the product is manufactured by a different firm. In other cases, design and manufacturing is accomplished by the same firm. Whatever the case, the product cycle begins with a concept, an idea for a product. This concept is cultivated, refined, analyzed, improved, and translated into a plan for the product through the design engineering process. The plan is documented by drafting a set of engineering drawings showing how the product is made and providing a set of specifications indicating how the product should perform.

Except for engineering changes, which typically follow the product throughout its life cycle, this completes the design activities. The next activities involve the manufacturing of the product. A process plan is formulated that specifies the sequence of production operations required to make the product. New equipment and tools must sometimes be acquired to produce the new product. Scheduling provides a plan that commits the company to the manufacturing of certain quantities of the product by certain dates. Once all of these plans are formulated, the product goes into production (which is, in the modern way of understanding, an interrelated process of manufacturing and quality checks).

2.6.2 The Integration of CAD and CAM

The impact of CAD / CAM is manifest in all of the different activities in the product cycle, as indicated in Fig. 2.8 [Groover and Zimmers, 1984]. Computer aided design and automated drafting are utilized in the conceptualization, design, and documentation of the product. Computers are used in process planning and scheduling in order to perform these functions more efficiently. Computers are used in production to monitor and control the manufacturing operations. In quality control, computers are used to perform inspections and performance tests on the production process and the product.

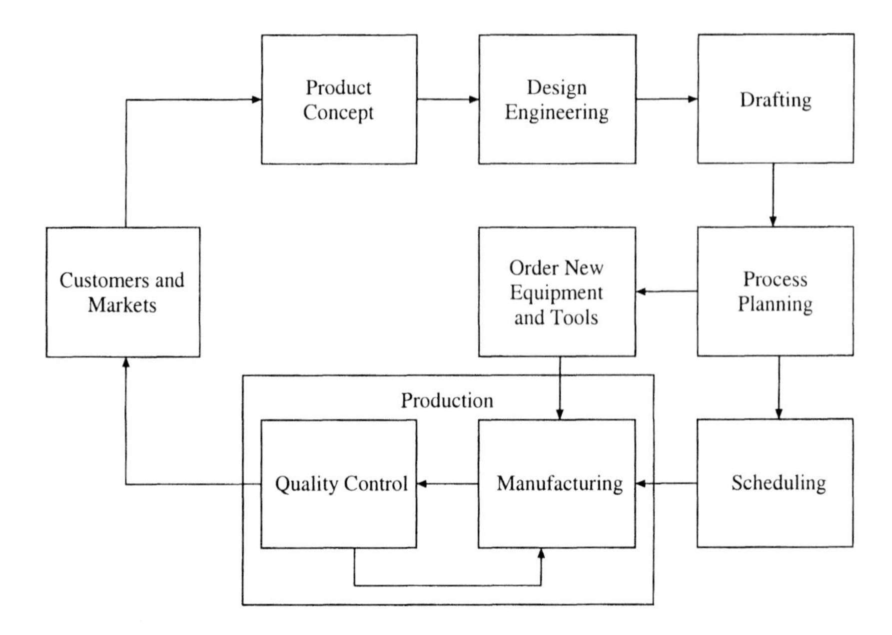

Figure 2.7: Product cycle: design and manufacturing

As illustrated in Fig. 2.8, CAD / CAM is overlaid on virtually all of the activities and functions of the product cycle. In the design and production operations of a modern manufacturing firm, the computer has become a pervasive, useful, and indispensable tool. It is strategically important and competitively imperative that manufacturing firms and the people who are employed by them understand CAD / CAM thoroughly.

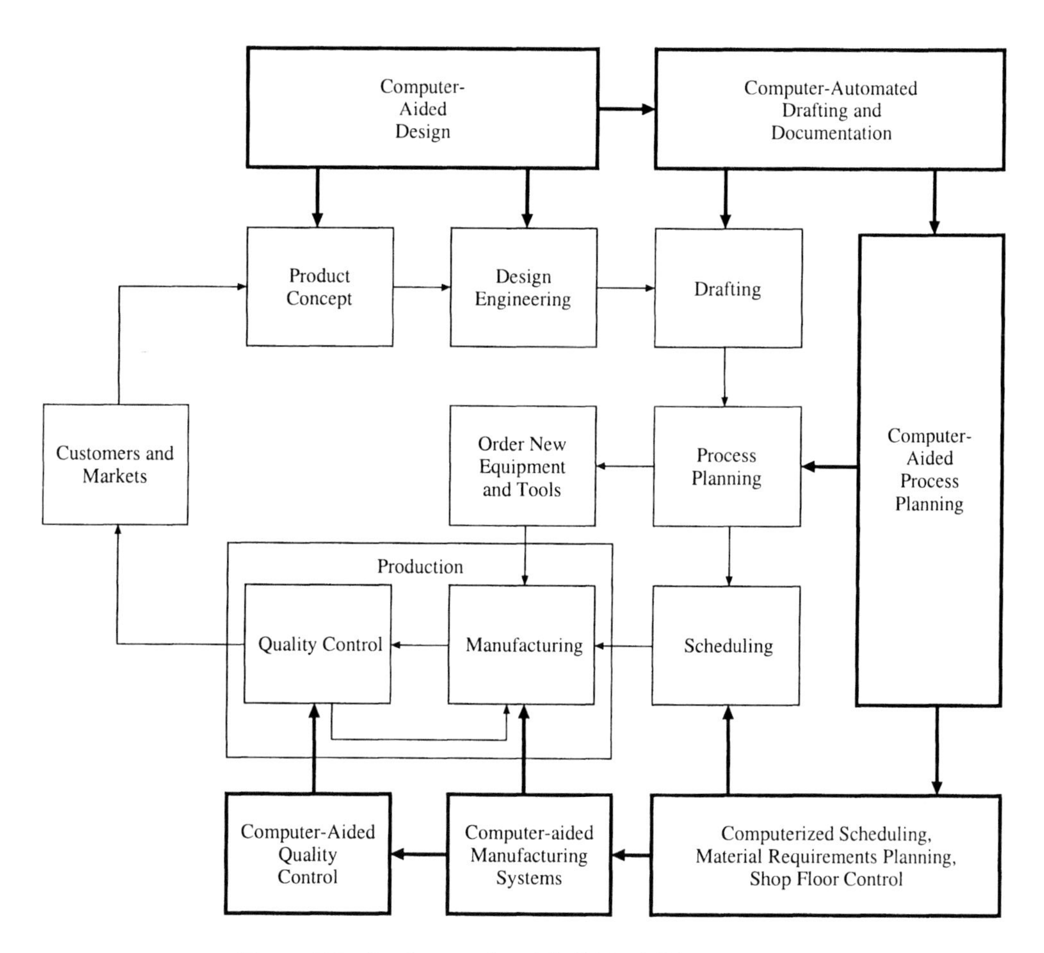

Figure 2.8: Product cycle with CAD / CAM support

Chapter 3

Computer Automated Manufacturing

This chapter informs about trends in factory automation and computer integrated manufacturing. The development of industrial robots and numerically controlled (NC) machines is sketched together with modern programming concepts in order to achieve a higher degree of automation and less production errors.

3.1 The Era of Factory Automation

Factory automation has become a widely used concept in industry, particularly since the end of the Eighties (see [Eversheim *et al.*, 1988]). The following trends can be observed when considering modern installations:

1. Even rather small companies with only few computer controlled mechanisms (robots, numerically controlled (NC) machines) introduce computer aided automation concepts into their plants.
2. An increasing number of companies begin to automate tasks with high degrees of complexity and diversification.
3. Heading towards a one-of-a-kind production, the lots get more and more reduced, whereas the product complexity increases.

The main reasons for the need of factory automation and flexible manufacturing, in order to make the production process more efficient, are:

- Shorter innovation periods imply faster production changes and smaller lots for a production series.
- Programming need not be done any longer on the NC machines, which implies no loss of production.
- Repetitive activities (e. g., positioning of workpieces and tools, see [Eversheim and Holz, 1982]) are reduced.

- Manufacturing errors can already be detected and overcome during the design stage.
- Due to the use of conventional NC machining techniques, the pressure of low-income countries is on the increase.
- Because of a higher flexibility, production can more easily be adapted to market changes.
- The programmers' potential is concentrated on the creative parts of the manufacturing process.

By standardizing assembly groups, an additional series effect can be reached (see [Jaissle, 1977]), which intensifies in special manufacturing type concepts and a better potential for automation (design for automation, design for (dis-)assembly, etc.).

In modern manufacturing, it has to be striven for high production flexibility regarding low production costs. High production flexibility allows a demand-adapted level of production with short start times and quick production flow times, a decrease of the circulating stock implying lower capital inactivation and, finally, a reduction of the costs per piece. Basic requirements of such a cost-saving flexible manufacturing are:

1. independence of expensive production devices, and
2. quickly adaptable processing machines and logistics.

For these reasons, the activities of factory automation are directed to automate process planning, too, and to integrate this step into systems for integrated, computer aided planning and production. Thus, a complete description of a planned task, including organizational information, has to supply information for the whole planning process and production process.

These constellations lead to production units and manufacturing cells that need to be reprogrammed rather frequently, but are too complicated to be programmed and tested by human experts only. Therefore, computer aids are in the focus of research today in manufacturing design, programming, simulation, and verification (see, e. g., [Chang *et al.*, 1997]).

3.2 Extending the Scheme of Computer Integrated Manufacturing

Computer integrated manufacturing (*CIM*) began to constitute a key topic and key goal in factory automation about fifteen years ago (see e. g., [Graiser, 1983], [Kochan, 1986]). However, for several years the engineering part of CIM was more or less restricted to *computer aided design* (*CAD*), *computer aided manufacturing* (*CAM*), and the data exchange ("integration") between these two C-techniques [Spur and Krause, 1984].

During the Nineties, steady improvements in computer performance as well as new research results in algorithm design and other fundamental fields (e. g., computational geometry, artificial intelligence) have opened the gate towards CIM considerably. Among others, the following C-techniques have entered the CIM-stage:

- **Computer aided testing (CAT)**:
  By means of graphic simulation and algorithmic verification of machining and manufacturing processes, a technologically sound production process shall be guaranteed that results in the desired product.

- **Computer aided quality control (CAQ)**:
  Given specific technological characteristics of production devices and material, a manufacturing process yielding the desired quality standards (i. e. best value-for-money ratio with respect to quality regarding technological restrictions) shall be guaranteed and checked.

- **Computer aided process planning (CAP)**:
  NC programming and manufacturing control should also be linked to computer aided process planning systems. The aim of this integration of CAD, CAP, and CAM under the key–term *computer aided engineering* (*CAE*) is to reduce the amount of data and the number of files to describe the process.

Additionally, the following technique has become closely interconnected with CIM:

- **Realistic image animation**:
  Due to the rising hardware power, especially in the fields of graphics, no less than a realistic, trick-film-style animation can nowadays fulfill the demands. The applications range from factory planning through employee training to product marketing.

Consequently, computer assistance in factory automation evolves from construction drafting and design (replacing the drafting boards) and feature-style NC programming (assisting humans by the transformation of manufacturing tasks into NC punched tapes) into powerful integrated machining and manufacturing automation environments. Data exchange between the various modules can be done via computer networks, and the ideas of the human designer expert can be highly automatically realized and transformed into factory layouts and production logic. The created manufacturing tasks are automatically checked for correctness and production efficiency thus avoiding factory standstills by eliminating or correcting human errors in object or task design.

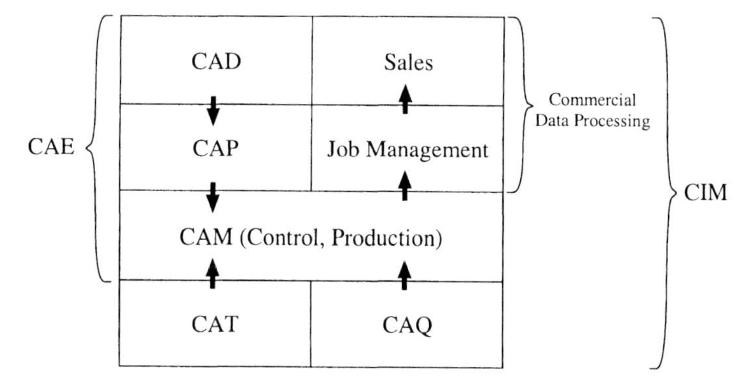

Figure 3.1: A compact scheme of CIM

A compact scheme of a modern way of looking at CIM is shown in Fig. 3.1.

3.3 History of Factory Automation

In this section, we give a brief review of the historical background of contemporary industrial robots and machining centers (work cells), which have their origins in both the teleoperator and the numerically controlled (NC) machine tool.

1. In 1947, the first servoed electric-powered teleoperator was developed. The teleoperator is a device to allow an operator to perform a task at a distance. It was developed during the second world war to handle radioactive materials, substituting the operator's hands [Goertz, 1963].

2. Early in the 1950s, the Servomechanisms Laboratory at the Massachusetts Institute of Technology (M.I.T.) developed the first NC milling machine for manufacturing advanced aircraft parts, initiated by the US Air Force. The research was to combine sophisticated servo system expertise with the newly developing computer techniques. Using a Whirlwind computer, the pattern to be cut was stored in digital form on a punched tape and then a servo- controlled milling machine cut the metal [Rosenberg, 1972]. It is an interesting fact that computer aided manufacturing is not a descendant of computer aided design, but has a distinct origin of its own.

3. The activities of the M.I.T. led to the evolution of the APT (Automatically Programmed Tools) language. The objective of the APT research was to provide a means by which the part programmer could input the machining instructions into the NC machine in simple English-like statements. Later, the step from APT to design programs including computer graphics functions was outlined by Coons.

4. In 1954, Denavit and Hartenberg published a very economic scheme for describing the kinematic structure of a robot or an NC machine [Denavit and Hartenberg, 1954]. Their approach allowed the treatment of originally mechanical problems, like positioning the gripper of a robot, using well-known concepts of classical geometry.

5. In the year 1958, Wiedemann Inc. in Pennsylvania, USA, presented the first NC turret punching machine. For the first time, tool changes were realized. This was done by rotating the turret magazine; a principle used more or less unchanged in turret stamping machines nowadays [Mayr, 1987].

6. In the 1960s, the first Unimate industrial robot was presented [Engelberger, 1980], a device combining the articulated linkage of a teleoperator with the servoed axes of an NC milling machine. The industrial robot could be taught to perform any simple job by driving it by hand through the sequence of task positions, which were recorded. Task execution consisted in replaying these positions by servoing the individual joint axes of the robot. The industrial robot was soon be found to be ideal for pick-and-place jobs such as feeding a conveyor with parts.

7. In 1965, Roberts introduced homogeneous transformations as a suitable data structure for the description of the relative position and orientation between objects [Roberts, 1965]. The remaining problem to solve was that a manipulator, stripped of its touch sensors, relied on position-servoed joint axes. Homogeneous transformations, however, expressed the position and orientation of the end effector in Cartesian coordinates, not as the angles between a chain of non-orthogonal manipulator joints. In [Pieper, 1968],

the theories of closed-link chains were applied to obtain a solution to this problem, and the manipulator could then be commanded to move to Cartesian positions in the workspace.

8. In the beginning of the 1970s, in Japan research led to a hand-eye system that could assemble block structures when presented with an assembly drawing. The drawing was first viewed, the materials surveyed, and then the required structure built [Ejiri *et al.*, 1972]. This strategy was one of the first realized concepts of factory automation.

9. In 1976, the first computer numerically controlled (CNC) machine was presented, which allowed to program an NC machine directly via a connected computer [Geiger, 1978]. This concept also enables a bi-directional data exchange (direct-numerical control – DNC) in order to perform calibration and similar tasks.

10. The first computer controlled industrial robot was developed by Cincinnati Milacron in 1976 [Hohn, 1976]. This robot was able to interact with a moving conveyor whose position was sensed by a digital encoder.

3.4 Modern Devices and Concepts in Factory Automation

3.4.1 Industrial Robots

The German Association of Engineers defines an *industrial robot* (short: *robot*, see Fig. 3.2) as a re-programmable handling device that can be utilized in various fields of application [Weseslindtner, 1984]. Generally, industrial robots are used for moving material, tools, and other specialized manufacturing components, in order to solve a certain manufacturing problem specified by a programmed task. These tasks can be subdivided into

- **handling of tools**, including applications like, e. g., spot welding, arc welding, painting, coating, deburring, grinding,
- **handling of workpieces**, including applications like, e. g., feeding, removal, applying fixtures, and
- **assembly**.

The computer controlled industrial robot represents the first truly general purpose automation device [Paul, 1981]. An industrial robot can be programmed to perform any number of jobs and will eliminate the need for high cost, custom designed automation equipment. It can provide for the automation of product assembly, and its low cost will make the automation of small batch production shops possible.

The increasing use of industrial robots can be seen in the frame of modern manufacturing technologies and the increasing level of automation. Within these fields, industrial robots offer a number of benefits. Typical examples are:

- In mass production the same robot type can be used for completely different tasks within one factory plant. Once programmed, the robot is able to perform the same task over a long time with a high reliability and an excellent productivity.

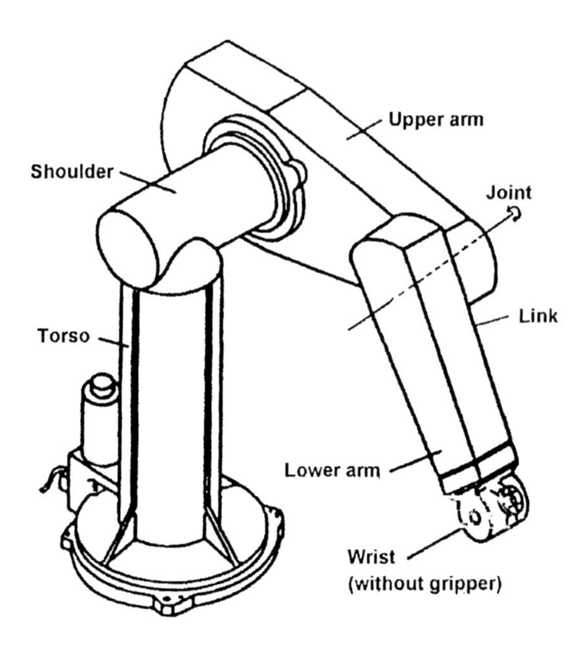

Figure 3.2: Sample industrial robot

- In low-quantity production the flexibility and re-programmability of the industrial robot is the key feature. If the programming of the robot is supported by suitable software, the same robot can be profitably used for a varying spectrum of tasks.

However, due to rather high down-times and expensive maintenance costs of the first robot installations, the enthusiasm of the early days has had a sobering effect on the industry. Still nowadays, where the down-time / use ratio of industrial robots is as good as that of NC machines, engineers are often skeptic when a robot shall be integrated into a manufacturing cell.

For the future one can expect an increasing number of installations of linked manufacturing systems. In these systems, several robots and similar devices (e. g., conveyor-belt feeders) interact with each other. The key problems to solve will be parallel cooperation of the devices and communication between the parallel processes.

3.4.2 Numerically Controlled (NC) Machines

Numerical control (*NC*) can be defined as a form of programmable automation in which the process is controlled by numbers, letters, and symbols. In NC, the numbers form a program of instructions designed for a particular workpiece or job. When the job changes, the program of instructions is changed. This capability to change the program for each new job is what gives NC its flexibility. It is much easier to write new programs than to make major changes in the production equipment [Groover and Zimmers, 1984].

NC machines (see Fig. 3.3; Courtesy of Anilam Electronics, Fa. Core, Düsseldorf, Germany) are widely used in industry today, especially in the metalworking industry. By far the most common application of NC is for metal cutting machining. Within this category, NC equipment has been built to perform virtually the entire range of material removal processes, including:

- milling,
- drilling,
- turning,
- grinding, and
- sawing.

Figure 3.3: Sample NC machine

Within the machining category, NC machine tools are appropriate for a broad variety of jobs. In the following, we list the general characteristics of production jobs in metal machining for which NC would be most appropriate:

- Parts are processed frequently and in small lots.
- The part geometry is complex.
- Many operations have to be performed on the part in its processing.
- Much metal needs to be removed.
- Changes in the engineering design are likely to take place, even after the production start.
- Close tolerances must be held on the workpiece.
- The parts require 100 % inspection.

It has been estimated that most manufactured parts are produced in lot sizes of 50 or fewer [Droy, 1981]. Small-lot and batch production jobs represent the ideal situations for the application of NC. This is made possible by the capability to program the NC machine and to save that program for subsequent use in future orders. If the NC programs are long and complicated (complex part geometry, many operations, much metal removed), this makes NC all the more appropriate when compared to manual methods of production. If engineering design changes or shifts in the production schedule are likely, the use of tape control provides the flexibility needed to adapt to these changes. Finally, if quality and inspection are important issues (close tolerances, high part costs, 100 % inspection required), NC would be most suitable, owing to its high accuracy and repeatability.

As already stated in Chapter 3.3, in technical practice, NC has been substituted by CNC and DNC, but in accordance to industrial name conventions we use NC as a summary for CNC, DNC, and similar control variants. The geometric and kinematic characteristics of NC machines will be discussed in more detail in Chapter 6.9.1. A detailed exposition of the concept of NC is presented in [Bézier, 1972]; an overview on numerical controllers is given in [Vossloh, 1983] and [Sautter, 1987].

3.4.3 Programming of Robots and NC Machines

Programmable material handling devices, such as industrial robots, open up new modes of automation for small and medium-size batch production (see, e. g., [Spur *et al.*, 1976], [Rahmacher and Heßelmann, 1983]). An economic utilization of such handling devices requires extensive planning work. In most cases, planning is based mainly on the individual experience of the planner, and therefore frequently leads to unsatisfactory results. Important tasks in planning the application of industrial robots are the determination of the number of required handling devices, the layout of the machine tool installation, the positioning of the handling devices, the determination of the optimal work paths, and the provision of control data. The application of more than one handling device results in the additional task of coordinating the number and sequence of the partial tasks to be performed by the individual robots. The aim when planning handling tasks for a machine tool installation with robots is to use as few motion axes as possible and to achieve short work paths with minimum energy.

The determination of optimal work paths depends on the following parameters:

- the geometric facts,

- the kinematics of the handling devices,
- the motion speed of the individual axes,
- the operating sequence, and
- the gripper design.

The computation of the optimal work path includes the task of providing the control data for the programmable handling device. Various programming tools are available for the programming of handling devices:

- With the *manual method* the manufacturing task is programmed by mechanically interconnecting the control circuit of the robot via a plug board of the handling device.
- With the *playback method* the robot is programmed by guiding its tool holder or gripper through its work path. The actual position values obtained hereby are recorded in defined time intervals and then stored. They produce the user program.
- Programming by *teach-in* is done by commanding the robot through its trajectory with the help of a teach unit that contains switches and push-buttons to move the robot.
- *Offline programming* is done with robot independent peripheral computing devices. Once the program is completed it is loaded into the control system of the robot.

Manual, playback, and teach-in methods are still rather widespread in robot programming. A modern version of these techniques is the facility of online textual programming. Here the task program is entered into the robot via a terminal using a special robot programming language. Testing and verification of the program, however, has still to be done directly on the robot.

Such a "programming on the shop floor" has several disadvantages:

- During the programming and testing of the task the robot cannot be used for production.
- For more complex tasks the programming may take days or even weeks, which means a considerable reduction of the production capability.
- Available product data (e. g., the geometry from a CAD system) cannot be used directly and has to be entered a second time. Thus, the possibility of input errors is magnified.
- Planning in advance (cycle time evaluation, process planning, etc.) is possible only in a very restricted way.

Due to the disadvantages of the shop floor programming by manual, playback, or teach-in methods, *offline programming* is gaining importance for the programming of industrial robots. Such systems have been used for the programming of NC machines for several years already; now many of these systems are extended by a module for the programming of industrial robots (at least for special types and applications).

The existence of uniform interfaces and defined basic functions are the preconditions for the application of universal offline programming methods [Rembold and Dillmann, 1986]. For a detailed analysis of robot / NC code standardization, we refer to [Mayr, 1991a].

In offline programming, two main groups of programming strategies have to be distinguished [VDI, Verein Deutscher Ingenieure, 1983]:

1. explicit or motion-oriented programming and
2. implicit world model or task-oriented programming.

In explicit programming, every motion of the robot for a task has to be described individually. The exact definition of the positions and orientations in three-dimensional space require the programmer to have an extraordinary imaginative capacity to view objects in this space, which cannot always be presupposed.

The advantage of the implicit, world model oriented languages over the explicit languages lies in the fact that the description of complicated motions and logical relations is omitted and programming is application-oriented. This is one of the main reasons why these kinds of languages are well suited for programming of interlinked manufacturing facilities. In particular, with assembly processes it is necessary to define program sequence instructions, error detection and correction routines, and monitoring functions by a problem oriented language.

Implicit programming languages should have the following features:

- description of the motion trajectory,
- description of the operation sequence,
- description of the workpiece and its environment, and
- monitoring and supervision.

In the field of NC programming, offline techniques led to the definition of machining paths by surface tracing, feature analysis, and similar strategies. A recent trend is to unify the online programming interfaces and the offline programming modules in order to achieve a common programming environment for offline task design and online task adaptation. This "workshop oriented programming" (WOP) shall lead to a better interaction between the offline NC programmer and the machining expert that supervises the NC production process online (see, e. g., [Gödde, 1990], [Wätzig and Cajar, 1990]).

Chapter 4

Analytical Geometry

This chapter introduces the mathematical background for positioning and visualizing objects. The analytical geometry approach is pursued using homogeneous coordinates and transformation matrices. All operations are performed in an orthonormal Euclidean coordinate system.

4.1 Coordinate Transformations and Their Realization

4.1.1 Rotation

As an explanatory example, we consider the rotation of the XY plane with respect to the Z axis, see Fig. 4.1. Let the rotation be in mathematical positive direction, i. e., counterclockwise. The new coordinates of an arbitrary point $\mathbf{P}$ are computed trigonometrically as sketched and can be written using the following notation:

Rotation (of the coordinate system):

$$\begin{pmatrix} x' \\ y' \\ z' \end{pmatrix} = \begin{bmatrix} \cos\alpha & \sin\alpha & 0 \\ -\sin\alpha & \cos\alpha & 0 \\ 0 & 0 & 1 \end{bmatrix} \begin{pmatrix} x \\ y \\ z \end{pmatrix}.$$

This *rotation matrix* R has to be *orthonormal* due to the orthonormality of the Euclidean coordinate system. This means that $RR^T = I$ (the matrix multiplied by its transposed yields the unit matrix). In other words,

- each row vector and each column vector must have length 1,
- the scalar multiplication of any row vector with any other row vector must be 0, and also
- the scalar multiplication of any column vector with any other column vector must be 0.[1]

[1] These conditions can be used for checking and "correcting" numerical errors of rotation matrices.

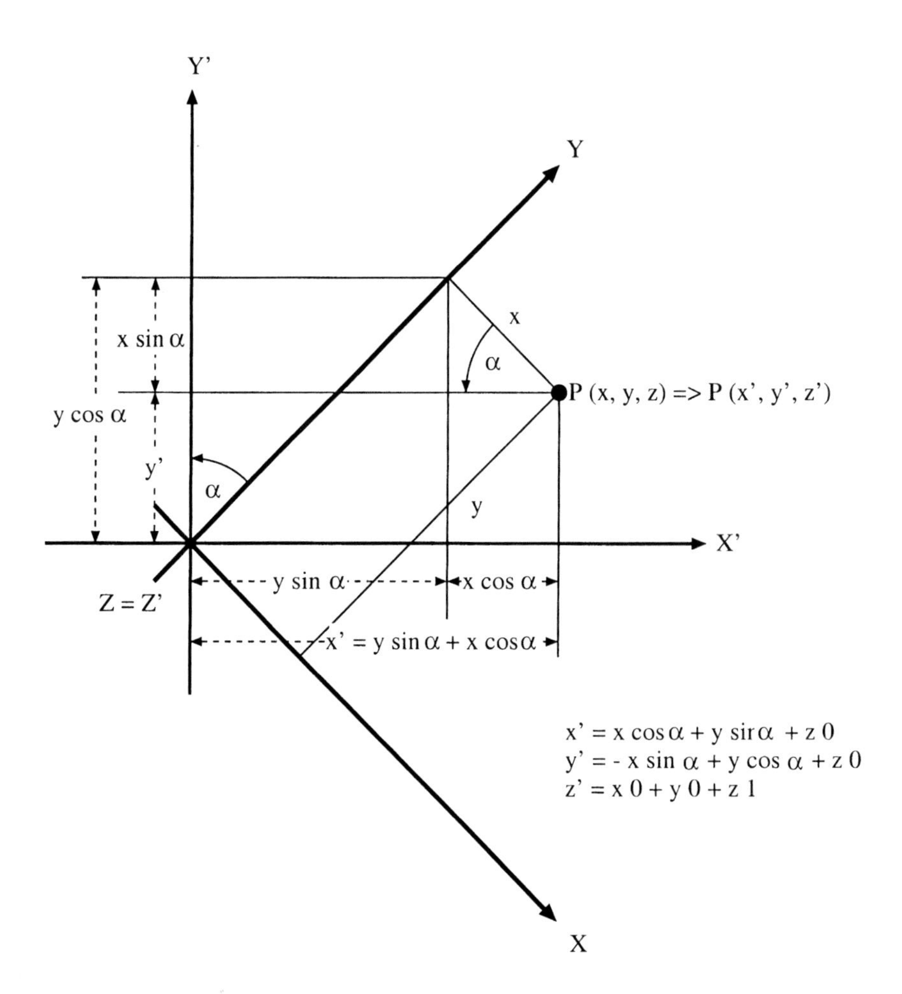

Figure 4.1: Rotating the coordinate system with respect to the Z axis

Since, in general, one does not want to rotate coordinate systems, but objects, α has to be replaced by $-\alpha$, resulting in the

Rotation (with respect to the Z-axis):

$$\begin{pmatrix} x' \\ y' \\ z' \end{pmatrix} = \begin{bmatrix} \cos\alpha & -\sin\alpha & 0 \\ \sin\alpha & \cos\alpha & 0 \\ 0 & 0 & 1 \end{bmatrix} \begin{pmatrix} x \\ y \\ z \end{pmatrix}.$$

Any arbitrary rotation in 3D can be described by a sequence of rotations with respect to the single coordinate axes. A rotation with respect to the Y (X) axis by an angle of β (γ) can be specified analogously in the following way:

Rotation (with respect to the Y-axis):

$$\begin{pmatrix} x' \\ y' \\ z' \end{pmatrix} = \begin{bmatrix} \cos\beta & 0 & \sin\beta \\ 0 & 1 & 0 \\ -\sin\beta & 0 & \cos\beta \end{bmatrix} \begin{pmatrix} x \\ y \\ z \end{pmatrix}.$$

Rotation (with respect to the X-axis):

$$\begin{pmatrix} x' \\ y' \\ z' \end{pmatrix} = \begin{bmatrix} 1 & 0 & 0 \\ 0 & \cos\gamma & -\sin\gamma \\ 0 & \sin\gamma & \cos\gamma \end{bmatrix} \begin{pmatrix} x \\ y \\ z \end{pmatrix}.$$

If an arbitrary rotation is composed of rotations with respect to the single axes, it is important to keep in mind that *matrix multiplication is not commutative*! Any order of the single rotations is, of course, valid, but the matrices have to be adapted accordingly. We fix the *arbitrary rotation in 3D* in the following way:

$$R_3 = R_Z R_Y R_X.$$

This order specifies that first a rotation is performed with respect to the X axis, then with respect to the Y axis, and finally with respect to the Z axis, i. e.:

$$\begin{pmatrix} x''' \\ y''' \\ z''' \end{pmatrix} = R_Z(R_Y(R_X \begin{pmatrix} x \\ y \\ z \end{pmatrix})) = R_Z R_Y R_X \begin{pmatrix} x \\ y \\ z \end{pmatrix} = R_3 \begin{pmatrix} x \\ y \\ z \end{pmatrix}.$$

Utilizing the associativity of matrix multiplication this way, one can compute the rotation (transformation) matrix just once and then apply it to all objects that are to be transformed (*preprocessing idea*). If the three rotation matrices are multiplied, one gets the *general rotation matrix in 3-space*:

$$R_3 = \begin{bmatrix} \cos\alpha\cos\beta & -\sin\alpha\cos\gamma + \cos\alpha\sin\beta\sin\gamma & \sin\alpha\sin\gamma + \cos\alpha\sin\beta\cos\gamma \\ \sin\alpha\cos\beta & \cos\alpha\cos\gamma + \sin\alpha\sin\beta\sin\gamma & -\cos\alpha\sin\gamma + \sin\alpha\sin\beta\cos\gamma \\ -\sin\beta & \cos\beta\sin\gamma & \cos\beta\cos\gamma \end{bmatrix}.$$

4.1.2 Reflection

In order to reflect an object with respect to an arbitrary 2D plane, one first rotates the scenario in such a way that one can perform the reflection with respect to a coordinate plane, e. g., the XY plane. Reflection with respect to the XY plane means changing the sign of the Z values, i. e., in matrix notation,

Reflection (with respect to the XY-plane):

$$\begin{pmatrix} x' \\ y' \\ z' \end{pmatrix} = \begin{bmatrix} 1 & 0 & 0 \\ 0 & 1 & 0 \\ 0 & 0 & -1 \end{bmatrix} \begin{pmatrix} x \\ y \\ z \end{pmatrix}.$$

For an arbitrary reflection, the inverse transformation of the scenario has to be applied after the reflection.

4.1.3 Scaling

Scaling is used for changing the dimension of objects. This is, e. g., necessary for adjusting the size of two objects that are to be adjoined. This transformation can be achieved in the following way:

Scaling (varying for each dimension):

$$\begin{pmatrix} x' \\ y' \\ z' \end{pmatrix} = \begin{bmatrix} S_X & 0 & 0 \\ 0 & S_Y & 0 \\ 0 & 0 & S_Z \end{bmatrix} \begin{pmatrix} x \\ y \\ z \end{pmatrix}.$$

S_X, S_Y, and S_Z are the scaling factors for each single dimension. In case of $S_X = S_Y = S_Z = S$, this kind of scaling is called *regular*. A more compact notation for regular scaling is presented in Chapter 4.2.2.

4.2 Introduction of Homogeneous Coordinates

The *homogeneous coordinate representation* of objects in an n-dimensional space is an $(n+1)$ dimensional entity such that a particular perspective projection recreates the original objects in n-space. This can be viewed as adding an extra coordinate, i. e. a scale factor. Hence, the 3D vector $(a, b, c)^T$ would be transformed into the homogeneous coordinate representation $(x, y, z, w)^T$, where $a = x/w$, $b = y/w$, $c = z/w$. Usually, the scale factor w equals 1. Vectors of the form $(x, y, z, 0)^T$ are used for representing points at infinity and directions. The vector $(0, 0, 0, 0)^T$ is undefined.

4.2.1 Translation

Using the concept of homogeneous coordinates, rotation as well as translation can be described by means of matrix multiplication. First, the general rotation matrix has to be expanded by one dimension thus taking homogeneous coordinates as arguments:

General Rotation (homogeneous coordinates):

$$\begin{pmatrix} x' \\ y' \\ z' \\ 1 \end{pmatrix} = \begin{bmatrix} R_3(1,1) & R_3(1,2) & R_3(1,3) & 0 \\ R_3(2,1) & R_3(2,2) & R_3(2,3) & 0 \\ R_3(3,1) & R_3(3,2) & R_3(3,3) & 0 \\ 0 & 0 & 0 & 1 \end{bmatrix} \begin{pmatrix} x \\ y \\ z \\ 1 \end{pmatrix}.$$

A sole translation can be formulated in the following way using homogeneous coordinates:

Translation (homogeneous coordinates):

$$\begin{pmatrix} x' \\ y' \\ z' \\ 1 \end{pmatrix} = \begin{pmatrix} x \\ y \\ z \\ 1 \end{pmatrix} + \begin{pmatrix} dx \\ dy \\ dz \\ 0 \end{pmatrix} = \begin{bmatrix} 1 & 0 & 0 & dx \\ 0 & 1 & 0 & dy \\ 0 & 0 & 1 & dz \\ 0 & 0 & 0 & 1 \end{bmatrix} \begin{pmatrix} x \\ y \\ z \\ 1 \end{pmatrix}.$$

These two transformations can now be combined into *one* transformation matrix with the semantics that *first the rotation* is performed, *then the translation* is added:

Combined Rotation and Translation:

$$\begin{pmatrix} x' \\ y' \\ z' \\ 1 \end{pmatrix} = \begin{bmatrix} R_3(1,1) & R_3(1,2) & R_3(1,3) & dx \\ R_3(2,1) & R_3(2,2) & R_3(2,3) & dy \\ R_3(3,1) & R_3(3,2) & R_3(3,3) & dz \\ 0 & 0 & 0 & 1 \end{bmatrix} \begin{pmatrix} x \\ y \\ z \\ 1 \end{pmatrix}.$$

4.2.2 Regular Scaling

The most common form of scaling is the regular resizing of an object with respect to all dimensions. Such a *regular scaling* by the factor S can be performed in the following way (using homogeneous coordinates):

Regular Scaling:

$$\begin{pmatrix} x' \\ y' \\ z' \\ 1 \end{pmatrix} = \begin{pmatrix} Sx \\ Sy \\ Sz \\ 1 \end{pmatrix} =_H \begin{pmatrix} x \\ y \\ z \\ \frac{1}{S} \end{pmatrix} = \begin{bmatrix} 1 & 0 & 0 & 0 \\ 0 & 1 & 0 & 0 \\ 0 & 0 & 1 & 0 \\ 0 & 0 & 0 & \frac{1}{S} \end{bmatrix} \begin{pmatrix} x \\ y \\ z \\ 1 \end{pmatrix}.^2$$

This scaling matrix can again be multiplied with the matrix for combined rotation and translation yielding:

[2] $=_H$ indicates that homogeneous vectors are considered "equal", if their orientation is the same, even if their length is different. This transformation is called *homogenization*.

Combined Rotation, Translation, and Scaling:

$$\begin{pmatrix} x' \\ y' \\ z' \\ 1 \end{pmatrix} = \begin{bmatrix} R_3(1,1) & R_3(1,2) & R_3(1,3) & dx \\ R_3(2,1) & R_3(2,2) & R_3(2,3) & dy \\ R_3(3,1) & R_3(3,2) & R_3(3,3) & dz \\ 0 & 0 & 0 & \frac{1}{S} \end{bmatrix} \begin{pmatrix} x \\ y \\ z \\ 1 \end{pmatrix},$$

where first the rotation is performed, then the translation is added, and finally the scaling is done.

4.2.3 Example: Combined Transformations

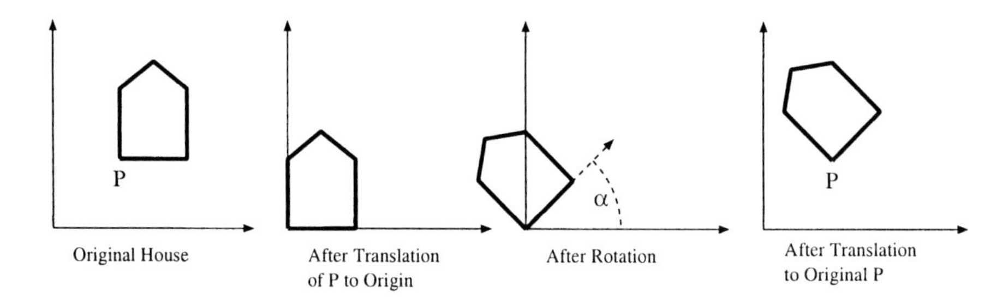

Figure 4.2: An example for combined transformations

Suppose that we want to rotate the simple house of Fig. 4.2 by an angle α, taking the point P_1 as the rotation center. This action is performed by splitting the problem into three fundamental transformations,

1. T_-: translating the center of rotation into the origin of the coordinate system,
2. R: rotating with respect to the Z axis (passing through the origin),
3. T_+ translating the center of rotation back to its starting position.

Using the matrix notation, this combined transformation can be described in the following way ($P_1 = (x_1\ y_1\ 0)$):

$$P' = T_+\ R\ T_-\ P, \quad i.e.,$$

$$\begin{pmatrix} x' \\ y' \\ z' \\ 1 \end{pmatrix} = \begin{bmatrix} 1 & 0 & 0 & x_1 \\ 0 & 1 & 0 & y_1 \\ 0 & 0 & 1 & 0 \\ 0 & 0 & 0 & 1 \end{bmatrix} \begin{bmatrix} \cos\alpha & -\sin\alpha & 0 & 0 \\ \sin\alpha & \cos\alpha & 0 & 0 \\ 0 & 0 & 1 & 0 \\ 0 & 0 & 0 & 1 \end{bmatrix} \begin{bmatrix} 1 & 0 & 0 & -x_1 \\ 0 & 1 & 0 & -y_1 \\ 0 & 0 & 1 & 0 \\ 0 & 0 & 0 & 1 \end{bmatrix} \begin{pmatrix} x \\ y \\ z \\ 1 \end{pmatrix}.$$

$$\begin{pmatrix} x' \\ y' \\ z' \\ 1 \end{pmatrix} = \begin{bmatrix} \cos\alpha & -\sin\alpha & 0 & x_1(1-\cos\alpha) + y_1\sin\alpha \\ \sin\alpha & \cos\alpha & 0 & y_1(1-\cos\alpha) - x_1\sin\alpha \\ 0 & 0 & 1 & 0 \\ 0 & 0 & 0 & 1 \end{bmatrix} \begin{pmatrix} x \\ y \\ z \\ 1 \end{pmatrix}.$$

This resulting matrix can now be applied to all elements of the object to be transformed.

4.3 The Concept of Coordinate Frames

The elements of a homogeneous transformation matrix can be interpreted as four vectors describing a seconding coordinate frame, relative to the original one. Consider the following transformation (as depicted in Fig. 4.3 [Paul, 1981]):

$$\begin{pmatrix} x' \\ y' \\ z' \\ 1 \end{pmatrix} = \text{Trans}(4,-3,7)\text{Rot}(y,90)\text{Rot}(z,90) = $$

$$= \begin{bmatrix} 0 & 0 & 1 & 4 \\ 1 & 0 & 0 & -3 \\ 0 & 1 & 0 & 7 \\ 0 & 0 & 0 & 1 \end{bmatrix} \begin{pmatrix} x \\ y \\ z \\ 1 \end{pmatrix}.$$

Obviously, the null vector $(0,0,0,1)^T$, i. e. the origin of the second frame, is transformed into $(4,-3,7,1)^T$. This gives the origin of the second frame with respect to the reference frame. The transformed unit vectors of the second frame are given by $(4,-2,7,1)^T$, $(4,-3,8,1)^T$, and $(5,-3,7,1)^T$. Subtracting the origin (and extending them to infinity), the vectors $(0,1,0,0)^T$, $(0,0,1,0)^T$, and $(1,0,0,0)^T$ are obtained. These direction vectors represent the axes of the second coordinate frame with respect to the reference frame. They correspond to the first three columns of the transformation matrix.

4.3.1 Relative Transformations

- If a product $A_n \cdot A_{n-1} \cdots A_1$ of homogeneous transformations A_i is *interpreted from right to left*, i. e. the transformation corresponding to A_{i+1} is executed after A_i, then each transformation is performed with *respect to the base reference coordinate system.* (See the example in Fig. 4.3: *Rot (Y, 90°)* refers to the Y-axis of the *base frame*!)

- If a product $A_n \cdot A_{n-1} \cdots A_1$ of homogeneous transformations A_i is *interpreted from left to right*, i. e. the transformation corresponding to A_i is executed after A_{i+1}, then transformation A_i is performed with *respect to the actual frame* described by the product $A_n \cdots A_{i+1}$.

4.3.2 Inverse Transformations

A homogeneous transformation T,

$$T = \begin{bmatrix} n_x & o_x & a_x & p_x \\ n_y & o_y & a_y & p_y \\ n_z & o_z & a_z & p_z \\ 0 & 0 & 0 & 1 \end{bmatrix},$$

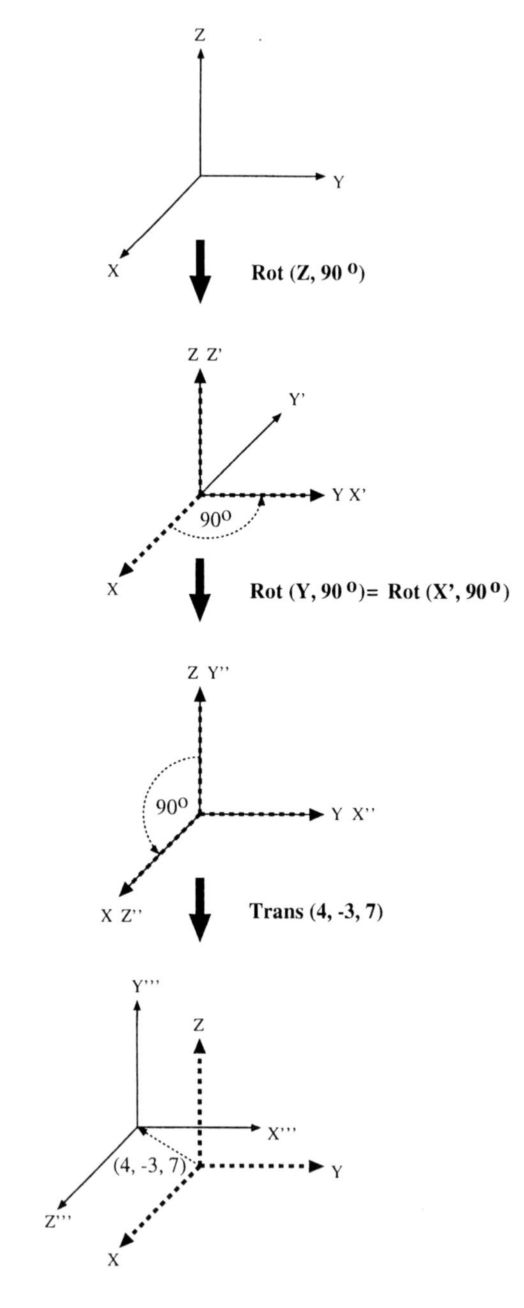

Figure 4.3: Interpreting a homogeneous transformation as a coordinate frame

is related to its inverse transformation T^{-1} by the formula

$$T^{-1} = \begin{bmatrix} n_x & n_y & n_z & -n \cdot p \\ o_x & o_y & o_z & -o \cdot p \\ a_x & a_y & a_z & -a \cdot p \\ 0 & 0 & 0 & 1 \end{bmatrix},$$

where p, n, o, a are the four column vectors and "$\cdot$" denotes the dot product.

This can be easily derived in the following way:

$$T = T_T.T_R = \begin{bmatrix} R_{3x3} & T_3 \\ 0\ 0\ 0 & 1 \end{bmatrix},$$

with

$$T_T = \begin{bmatrix} 1 & 0 & 0 & \\ 0 & 1 & 0 & T_3 \\ 0 & 0 & 1 & \\ 0 & 0 & 0 & 1 \end{bmatrix}, T_R = \begin{bmatrix} & & & 0 \\ & R_{3x3} & & 0 \\ & & & 0 \\ 0 & 0 & 0 & 1 \end{bmatrix}.$$

Since

$$T^{-1} = (T_T.T_R)^{-1} = T_R^{-1}.T_T^{-1},$$

with

$$T_T^{-1} = \begin{bmatrix} 1 & 0 & 0 & \\ 0 & 1 & 0 & -T_3 \\ 0 & 0 & 1 & \\ 0 & 0 & 0 & 1 \end{bmatrix}, T_R^{-1} = \begin{bmatrix} & & & 0 \\ & R_{3x3}^T & & 0 \\ & & & 0 \\ 0 & 0 & 0 & 1 \end{bmatrix}.$$

we get

$$T^{-1} = \begin{bmatrix} R_{3x3}^T & -R_{3x3}^T.T_3 \\ 0\ 0\ 0 & 1 \end{bmatrix}.$$

□

4.3.3 General Rotations

We briefly derive the transformation matrix representing a rotation by an angle θ about an arbitrary vector k passing through the origin: Let C be a coordinate frame,

$$C = \begin{bmatrix} n_x & o_x & k_x & p_x \\ n_y & o_y & k_y & p_y \\ n_z & o_z & k_z & p_z \\ 0 & 0 & 0 & 1 \end{bmatrix},$$

with z-axis unit vector equal to the (normalized) vector k. Rotating about the vector k is then equivalent to rotating about the z-axis of the frame C, $Rot(k, \theta) = Rot({}^C z, \theta)$. Imagine that the rotation about k is carried out with respect to the coordinate frame C. This yields the well-known transformation matrix

$$\mathrm{Rot}({}^C z, \theta) = \begin{bmatrix} \cos\theta & -\sin\theta & 0 & 0 \\ \sin\theta & \cos\theta & 0 & 0 \\ 0 & 0 & 1 & 0 \\ 0 & 0 & 0 & 1 \end{bmatrix}.$$

Hence, the transformation matrix representing the rotation by an angle θ about the vector k is given by

$$\mathrm{Rot}(k, \theta) = C \cdot \begin{bmatrix} \cos\theta & -\sin\theta & 0 & 0 \\ \sin\theta & \cos\theta & 0 & 0 \\ 0 & 0 & 1 & 0 \\ 0 & 0 & 0 & 1 \end{bmatrix} \cdot C^{-1}.$$

By expanding the right-hand side of this equation we obtain

$$\mathrm{Rot}(k, \theta) = \begin{bmatrix} k_x k_x \mathrm{c}^{-}\theta + \cos\theta & k_y k_x \mathrm{c}^{-}\theta - k_z \sin\theta & k_z k_x \mathrm{c}^{-}\theta + k_y \sin\theta & 0 \\ k_x k_y \mathrm{c}^{-}\theta + k_z \sin\theta & k_y k_y \mathrm{c}^{-}\theta + \cos\theta & k_z k_y \mathrm{c}^{-}\theta - k_x \sin\theta & 0 \\ k_x k_z \mathrm{c}^{-}\theta - k_y \sin\theta & k_y k_z \mathrm{c}^{-}\theta + k_x \sin\theta & k_z k_z \mathrm{c}^{-}\theta + \cos\theta & 0 \\ 0 & 0 & 0 & 1 \end{bmatrix},$$

where $\mathrm{c}^{-}\theta$ stands for $1 - \cos\theta$.

4.3.4 Finding Angle and Axis of Rotation

Suppose, we are given the matrix R of a homogeneous transformation and are to find an axis k which an equivalent rotation θ is made about, i. e. we are to solve the following equation for k and θ:

$$R = \begin{bmatrix} n_x & o_x & a_x & p_x \\ n_y & o_y & a_y & p_y \\ n_z & o_z & a_z & p_z \\ 0 & 0 & 0 & 1 \end{bmatrix} = \mathrm{Rot}(k, \theta).$$

Solving the equation

$$\begin{bmatrix} n_x\, o_x\, a_x\, p_x \\ n_y\, o_y\, a_y\, p_y \\ n_z\, o_z\, a_z\, p_z \\ 0\ 0\ 0\ 1 \end{bmatrix} = \begin{bmatrix} k_x k_x \mathrm{c}^-\theta + \cos\theta & k_y k_x \mathrm{c}^-\theta - k_z \sin\theta & k_z k_x \mathrm{c}^-\theta + k_y \sin\theta & 0 \\ k_x k_y \mathrm{c}^-\theta + k_z \sin\theta & k_y k_y \mathrm{c}^-\theta + \cos\theta & k_z k_y \mathrm{c}^-\theta - k_x \sin\theta & 0 \\ k_x k_z \mathrm{c}^-\theta - k_y \sin\theta & k_y k_z \mathrm{c}^-\theta + k_x \sin\theta & k_z k_z \mathrm{c}^-\theta + \cos\theta & 0 \\ 0 & 0 & 0 & 1 \end{bmatrix}$$

we obtain

$$\tan\theta = \frac{\sqrt{(o_z - a_y)^2 + (a_x - n_z)^2 + (n_y - o_x)^2}}{n_x + o_y + a_z - 1},$$

which uniquely defines θ for $0^o \leq \theta \leq 180^o$, and the equations

$$k_x = \mathrm{sgn}\ (o_z - a_y)\sqrt{\frac{n_x - \cos\theta}{1 - \cos\theta}},$$

$$k_y = \mathrm{sgn}\ (a_x - n_z)\sqrt{\frac{o_y - \cos\theta}{1 - \cos\theta}},$$

$$k_z = \mathrm{sgn}\ (n_y - o_x)\sqrt{\frac{a_z - \cos\theta}{1 - \cos\theta}}.$$

4.4 Projections

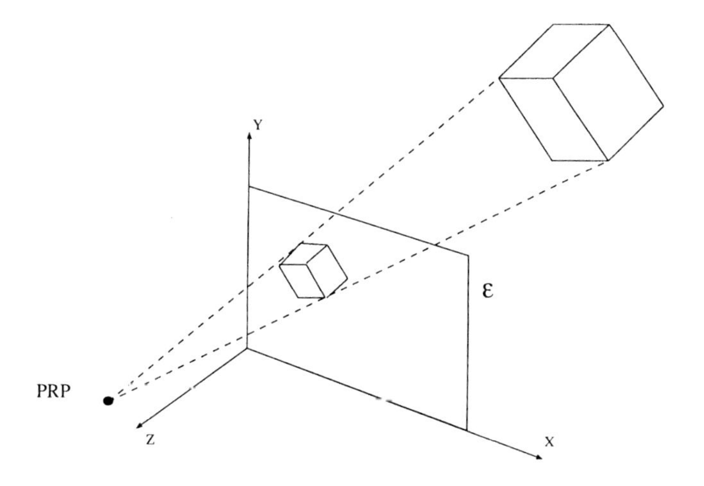

Figure 4.4: Projecting an object onto the projection plane

We can address the problem of projection in the convenient (X, Y, Z) coordinate system. We assume that our viewer is located at a point called the *projection reference point* (PRP). In Fig. 4.4, we see the view plane ε, points on an object, and projectors.

This type of projection, where all projectors emanate from a common point called the *center of projection*, is known as a *perspective projection*. It is the type of projection that is present in most physical image-formatting processes, such as the one used by the eye or camera. If we let the projection reference point move to infinity, the projectors become parallel to one another and define a *direction of projection*. This type of projection is naturally called a *parallel projection*.

Looking back at our three-dimensional viewing procedures, we see that there are only two fundamental types of viewing: parallel and perspective. On the other hand, classical graphics appears to have a host of different views, ranging from multi-view orthogonal projections to one-, two- and three-point perspective. This seeming discrepancy arises in classical graphics due to the desire to show a specific relationship among an object, the viewer, and the projection plane, as opposed to the computer graphics approach of complete independence of all specifications.

For example, when we draw an image of a house, we know which side we wish to display and thus where we should place the viewer in relationship to the building. Each classical view is determined by one such relationship; in computer graphics we usually have no coupling between the parameters to the functions that determine such a view.

In classical viewing, there is the underlying notion of a principal face. The types of objects viewed in real-world applications, such as architecture, tend to be composed of a number of planar faces, each of which can be thought of as a principal face. For a rectangular object, such as a house, there is natural notion of the front, back, top, bottom, right and left faces. In addition, since in many real-world objects the faces meet at right angles, such objects often have three orthogonal directions associated with them.

4.4.1 Orthogonal (Parallel) Projection

In the following, we will express points in view reference coordinates as (x, y, z), where x is the distance along the X axis, y along Y, and z along Z.

In the X, Y, Z system, the projection plane is the plane $z = 0$. Although it appears that the view-orientation transformation is adding an extra level of transformation and the additional floating-point operations to carry it out, this is not the case. The work is recovered by the simplicity of the project, and we even save work as subsequent clipping steps are simplified.

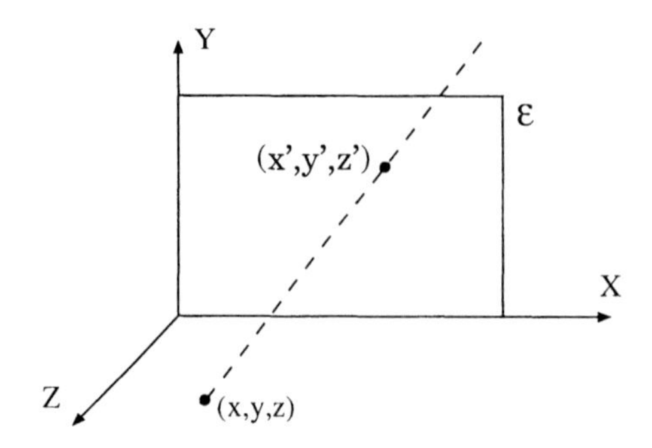

Figure 4.5: Orthogonal projection

In our coordinate system, an orthogonal projection is simple, as we can see from Fig. 4.5. In normalized projection coordinates, since the projection plane is the plane $Z = 0$, the projectors, being orthogonal to this plane, are lines of constant X and Y. If the operation of projection takes a point (x, y, z) and projects it into the point (x', y', z') in the projection plane, then, for our orthogonal projection, is $x' = x, y' = y$ and $z' = 0$.

This can be expressed using homogeneous coordinates. If p is the homogeneous coordinate representation of (x, y, z) and p' is its projection, then

$$p' = M_o p,$$

where M_0 is the *orthogonal projection matrix*

$$M_0 = \begin{bmatrix} 1 & 0 & 0 & 0 \\ 0 & 1 & 0 & 0 \\ 0 & 0 & 0 & 0 \\ 0 & 0 & 0 & 1 \end{bmatrix}.$$

This matrix allows us to view the act of projection as another transformation. If clipping were unnecessary, we could concatenate this matrix with the other matrices necessary for display purposes.

This matrix is not invertible. The non-singularity results from the fact that all points on a projector (i. e., points with the same z) project to the same point.

4.4.2 Perspective Projection

We can employ similar arguments with perspective viewing. First, the view orientation matrix will place the projection plane ε at $Z = 0$. Then, we can convert any perspective view to an orthogonal view by distorting the objects by a *perspective transformation*. We shall consider only the simple case, where the center of projection forms a right pyramid with the projection plane. Not only does this case correspond to most real-world viewing, but also, if we can solve this problem, we can solve the general case without much difficulty.

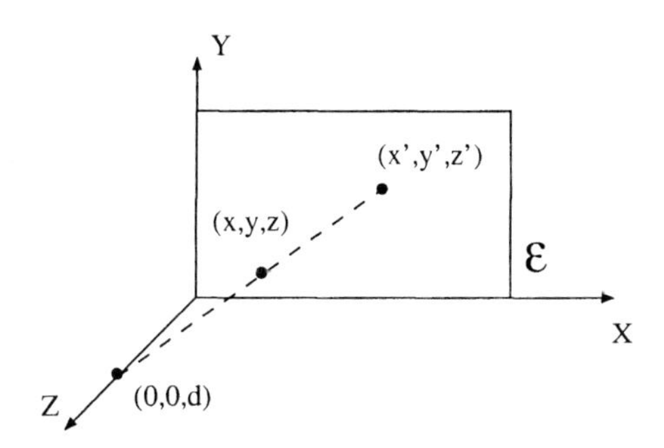

Figure 4.6: Perspective projection

Since the first steps are identical in parallel and perspective viewing, we can start in viewing reference coordinates. The viewing transformation places the projection plane at $Z = 0$

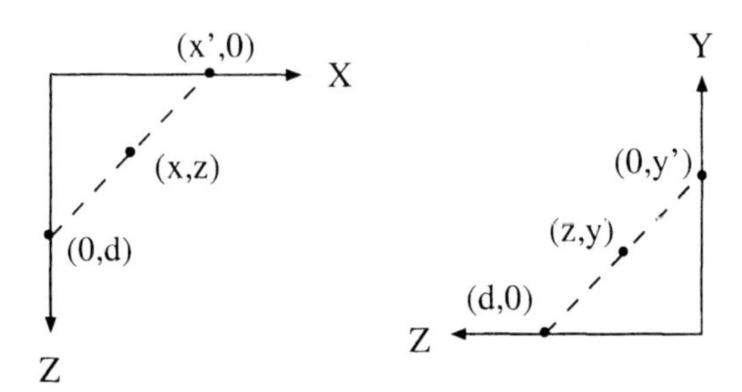

Figure 4.7: Top and side views of a perspective projection

and, for the simple perspective view, the center of projection is at $(0, 0, d)$, where $d > 0$. A projector connects this point with a point (x, y, z) on the object. These relationships are shown in Fig. 4.6. We can derive the equations of projection by considering the top and side views in Fig. 4.7. Solving for the projection point $(x', y', 0)$, we find

$$x' = \frac{x}{1 - z/d},$$
$$y' = \frac{y}{1 - z/d}.$$

The appearance of z in the denominator shows why objects farther from the viewer are smaller in perspective projections and also shows the nonlinearity of perspective viewing. We can also express this relationship using homogeneous coordinates. Consider the equation

$$p' = M_p p,$$

where M_p is the perspective matrix defined by the combination of a perspective transformation and a projection, i. e.

$$M_p = \begin{bmatrix} 1 & 0 & 0 & 0 \\ 0 & 1 & 0 & 0 \\ 0 & 0 & 0 & 0 \\ 0 & 0 & 0 & 1 \end{bmatrix} \begin{bmatrix} 1 & 0 & 0 & 0 \\ 0 & 1 & 0 & 0 \\ 0 & 0 & 1 & 0 \\ 0 & 0 & -\frac{1}{d} & 1 \end{bmatrix} = \begin{bmatrix} 1 & 0 & 0 & 0 \\ 0 & 1 & 0 & 0 \\ 0 & 0 & 0 & 0 \\ 0 & 0 & -\frac{1}{d} & 1 \end{bmatrix}.$$

These equations are deceptively simple. Suppose we write both p and p' as

$$p = \begin{bmatrix} x \\ y \\ z \\ w \end{bmatrix}, \qquad p' = \begin{bmatrix} x' \\ y' \\ z' \\ w' \end{bmatrix}.$$

These equations give us $x' = x$, $y' = y$, $z' = 0$ and $w' = w - \frac{z}{d}$. We get the standard perspective equations when we divide x', y' and z' by w'.

All operations – except the division by w' – can be performed using one transformation matrix when homogeneous coordinates are used. This is of great help for the actual implementation.

4.5 General Transformation Matrix

Various transformations in the field of analytical geometry have been described using homogeneous coordinates and transformation matrices. A general transformation matrix consists of the following parts:

General Transformation Matrix:

$$\begin{pmatrix} x' \\ y' \\ z' \\ 1 \end{pmatrix} = \begin{bmatrix} R_3(1,1) & R_3(1,2) & R_3(1,3) & T_x \\ R_3(2,1) & R_3(2,2) & R_3(2,3) & T_y \\ R_3(3,1) & R_3(3,2) & R_3(3,3) & T_z \\ P_x & P_y & P_z & \frac{1}{S} \end{bmatrix} \begin{pmatrix} x \\ y \\ z \\ 1 \end{pmatrix},$$

where R denotes the rotation matrix, T the translation vector, P the projection vector, and S the (regular) scaling factor. In order to reduce the variable components, without loss of generality the regular scaling factor S is normally set to 0, thus expressing scaling using the rotation matrix as shown in Chapter 4.1.3.

The following classes of transformations can be distinguished:

- **Linear transformations:**
  Linear transformations are characterized by $R \neq 0$, $T = 0$, and $P = 0$ in the general case.

  Possible transformations are rotation, scaling, and shearing (being a combination of the former two).

- **Affine transformations:**
  Affine Transformations are characterized by $R \neq 0$, $T \neq 0$, and $P = 0$ in the general case.

  They constitute a combination of linear transformations with translations. Affine transformations are closed with respect to combinations. An example for affine transformations are *orthogonal projections*, which show the following characteristic features:

  - The views are rather non-realistic (with respect to human experience). However, some important properties and measures remain unchanged.
  - The center of projection is shifted to infinity. Consequently, the rays of projection are parallel, the projection direction can be specified by a single vector.
  - Parallel lines remain parallel.
  - Measures can be extracted from the projection (regarding the respective scaling factors); angles remain unchanged only in the planes that are parallel to the projection plane.

- **Perspective transformations:**
  Perspective Transformations are characterized by $R \neq 0$, $T \neq 0$, and $P \neq 0$ in the general case.

  They constitute a combination of affine transformations with (non-affine) projections. An example for projective transformations are *perspective projections*, which show the following characteristic features:

- "Realistic" images can be generated that are comparable with the viewing capabilities of the human visual system (in 2D).
- The center of projection is placed in finite space ("eye point").
- The concept of perspective reduction is applied, reducing the size of objects with increasing distance from the eye point.
- Parallel lines, in general, do not remain parallel.
- Measures and angles, in general, can not be extracted from the projection.
- Parallel lines (that are not parallel to the projection plane) meet in "vanishing points".
- *Principal vanishing points* are vanishing points of lines that are parallel to a coordinate axis. The number of principal vanishing points (≤ 3 in 3-space; characterized by the number of non-zero components of the projection vector P) determines the type of perspective projection ("1 (2, 3) point projection").

4.6 Example: Generation of a Perspective Image

The following example is comprised of a combination of several of the concepts introduced throughout this chapter.

4.6.1 Problem Description

Given:

A block B defined by $x \in [1,4]$; $y \in [0,10]$; $z \in [1.5,3.5]$,
a projection center $CP = \left(-\sqrt{50}, 1.5, -\sqrt{50}\right)$,
a projection plane $P : z = -x$.

Find:

The perspective view of B with respect to CP and P.

The top view of this problem is depicted in Fig. 4.8.

4.6.2 Step-by-Step Approach

The problem of viewing B appropriately can be subdivided into the following steps:

- rotation of the projection plane with respect to the Y axis by an angle $\alpha = -45^o$ yielding the XY plane (i. e., $z = 0$),
- translation of the projection center into the XZ plane (i. e., $y = 0$),
- perspective projection onto the XY plane.

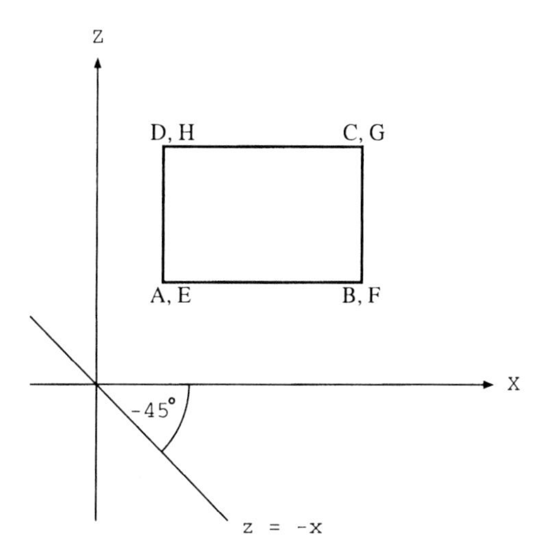

Figure 4.8: Transformation example: top view

4.6.3 Solution

The step-by-step approach results in the following solution:

- rotation: $M_R = R_Y\left(-45^o\right)^3$,
- translation: $M_T = T\left(0, -1.5, 0\right)$,
- perspective projection: $d = CP'_z = -10$; $M_P = P\left(-\frac{1}{d}\right) = P\left(0.1\right)$.

This yields the following combined transformation matrix:

$$M = M_P\ M_T\ M_R = \begin{bmatrix} 1 & 0 & 0 & 0 \\ 0 & 1 & 0 & 0 \\ 0 & 0 & 0 & 0 \\ 0 & 0 & 0.1 & 1 \end{bmatrix} \begin{bmatrix} 1 & 0 & 0 & 0 \\ 0 & 1 & 0 & -1.5 \\ 0 & 0 & 1 & 0 \\ 0 & 0 & 0 & 1 \end{bmatrix} \begin{bmatrix} 0.7 & 0 & -0.7 & 0 \\ 0 & 1 & 0 & 0 \\ 0.7 & 0 & 0.7 & 0 \\ 0 & 0 & 0 & 1 \end{bmatrix},$$

i. e.,

$$M = \begin{bmatrix} 0.7 & 0 & -0.7 & 0 \\ 0 & 1 & 0 & -1.5 \\ 0 & 0 & 0 & 0 \\ 0.07 & 0 & 0.07 & 1 \end{bmatrix}.$$

Thus, the projection is two-point perspective with the *vanishing points*

[3] This rotation additionally transforms the projection center into $CP' = (0, 1.5, -10)$.

$$VP_X = M \begin{pmatrix} 1 \\ 0 \\ 0 \\ 0 \end{pmatrix} = \begin{pmatrix} 0.7 \\ 0 \\ 0 \\ 0.07 \end{pmatrix} =_H \begin{pmatrix} 10 \\ 0 \\ 0 \\ 1 \end{pmatrix}$$

and

$$VP_Z = M \begin{pmatrix} 0 \\ 0 \\ 1 \\ 0 \end{pmatrix} = \begin{pmatrix} -0.7 \\ 0 \\ 0 \\ 0.07 \end{pmatrix} =_H \begin{pmatrix} -10 \\ 0 \\ 0 \\ 1 \end{pmatrix}.$$

By M, the given block B is transformed in the following way:

$$M \begin{pmatrix} 1 & 4 & 4 & 1 & 1 & 4 & 4 & 1 \\ 0 & 0 & 0 & 0 & 10 & 10 & 10 & 10 \\ 1.5 & 1.5 & 3.5 & 3.5 & 1.5 & 1.5 & 3.5 & 3.5 \\ 1 & 1 & 1 & 1 & 1 & 1 & 1 & 1 \end{pmatrix} =$$

$$= \begin{pmatrix} -0.35 & 1.75 & 0.35 & -1.75 & -0.35 & 1.75 & 0.35 & -1.75 \\ -1.5 & -1.5 & -1.5 & -1.5 & 8.5 & 8.5 & 8.5 & 8.5 \\ 0 & 0 & 0 & 0 & 0 & 0 & 0 & 0 \\ 1.175 & 1.385 & 1.525 & 1.315 & 1.175 & 1.385 & 1.525 & 1.315 \end{pmatrix} =_H$$

$$=_H \begin{pmatrix} -0.3 & 1.26 & 0.23 & -1.33 & -0.3 & 1.26 & 0.23 & -1.33 \\ -1.28 & -1.08 & -0.98 & -1.14 & 7.23 & 6.14 & 5.57 & 6.46 \\ 0 & 0 & 0 & 0 & 0 & 0 & 0 & 0 \\ 1 & 1 & 1 & 1 & 1 & 1 & 1 & 1 \end{pmatrix}.$$

This perspective view of B is depicted in Fig. 4.9.

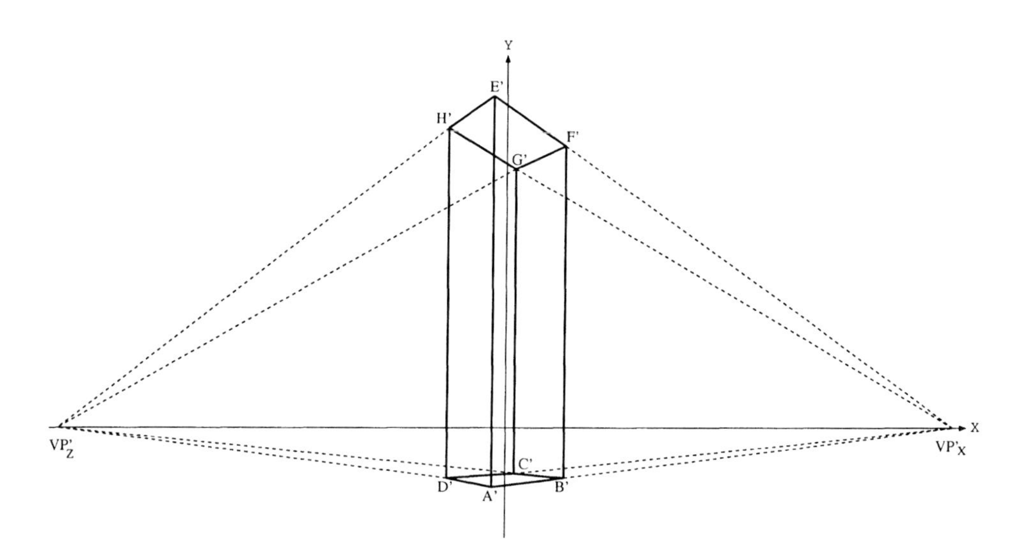

Figure 4.9: Transformation example: perspective view

Part II

Modeling

Chapter 5

Object Representation

The representation of 3D objects is discussed in this chapter, together with appropriate definitions and several alternatives for a classification of the different representations. This chapter forms a structured basis for Ch. 6.

5.1 Goals of Geometric Modeling

5.1.1 Basic Tasks to be Performed

Dealing with geometric aspects of the real world, one of the major problems one has to face is the suitable representation of geometric data (see geometric data processing, Ch. 2.1). Thus, the overall goal of geometric modeling is to represent geometric objects and entities in the computer such that some basic tasks and queries can be efficiently performed.

The class of user-requested operations heavily depends on the intended applications. However, the following operations should be applicable to most 3D models:

- answers to queries about the geometric model which are carried out through additional geometric data processing,
- creation of additional geometric entities needed for the application and derived from geometric entities defining the model,
- changing the geometric model data representation in order to meet specific application requirements.

In more detail, a representation scheme should support the following tasks [Encarnaç
ao and Schlechtendahl, 1983]:

- constructing and modifying a model,
- generating a projective display (with hidden-line elimination) for line-drawing hardware,
- producing technical drawings of the object (i. e. handling elevation, ground plane, etc.),

- visualizing an interesting object (with hidden surfaces removed) on area drawing devices,
- identifying geometric entities (such as edges or faces) in the 3D model by pointing to their 2D representations on a display,
- building new solids by means of set-theoretical operations,
- detecting and avoiding collisions,
- preparing data for NC machining,
- automated assembly,
- deciding inclusion / exclusion questions (i. e. is a given query point in the interior or in the exterior of the object?),
- computing geometric and inertial properties of objects (such as surface area, volume, mass, center of mass, moments of inertia, etc.),
- producing input data for other 'computer aided ...' systems (such as CAP, FEM analysis, etc.).

5.1.2 Specific Requirements for Factory Automation Applications

Since CAM applications are used to plan for and support manufacturing the part, they heavily depend on the model definition data. Thus, there is a strong link between geometric modeling and CAM applications which if strengthened and automated would result in a more sophisticated integration of CAD and CAM. An effective interface between geometric modelers and CAM applications is therefore needed and should provide the following [Elgabry, 1984]:

- geometric definition of the finished part containing information on surface finish and tolerances,
- geometric definition of the raw material (such as a casting, forging, bar or sheet stock, for instance),
- geometric definition of the cutting tool (Usually, this is a simplified representation covering critical geometric properties. It may include the tool holder geometry if considered critical.),
- geometric definition of the holding fixture,
- offsetting part geometry to arrive at the tool path definition points,
- volume decomposition for identifying rough cuts,
- representation of in-process part geometry (intermediate shapes during the NC process),
- calculation of the actual volume removal rate for each motion step,
- defining a tool swept volume for the purpose of subtracting it from the raw material in order to achieving the resulting geometry after machining,

- comparing the geometry resulting from machining with the geometry of the finished part (as it has been designed) in order to detect zones of over- or undercutting,
- detecting occurrences of collisions between the cutting tool and the fixture or machine tool, or between the tool and sections in the workpiece not meant to be machined.

5.2 Elementary Requirements for Representing 3D Objects

5.2.1 What is a Solid?

Although widely used, the term 'solid' is not strictly defined. Research on solid modeling has been initiated in the mid-seventies. Two of the founders were Braid [Braid, 1973] and Baumgart [Baumgart, 1974]. Requicha has suggested that the notion of an abstract solid – a subset of 3D Euclidean Space $\mathbf{E}^3$ – should mathematically capture the following properties (as stated in [Requicha, 1977, Requicha, 1980]):

- **Rigidity**: Shape properties of an abstract solid should be independent of its location and orientation.
- **Homogeneous three-dimensionality**: A solid must have an interior and its boundary must not contain 'dangling' (i. e. isolated) edges or faces.
- **Finiteness**: A solid should only occupy a finite portion of space.
- **Finite describability**: One should be able to store a solid using only a finite amount of storage space.
- **Closure under certain operations**: The result of applying rigid motions (such as translations and rotations) or Boolean operations for adding / removing material (such as welding, machining) to a solid should be a solid.
- **Boundary determinism**: The boundary of a solid should unambiguously determine its interior.

As a consequence, in [Requicha, 1977] it has been argued that a subset of $\mathbf{E}^d$ should be called a solid if it is

- bounded – i. e. definable within a finite portion of space,
- closed – most objects of the real world are closed,
- regular – an object S is regular if S is equal to the closure of its interior,
- semi-analytic – i. e. there is no pathological boundary behavior,

and if its surfaces are

- orientatable – i. e. the solid has a distinct interior,
- non-self-intersecting – the Klein bottle is a famous example of a self-intersecting, closed, and non-orientatable surface.

The orientatability of a closed polyhedral surface can be checked by applying the Möbius rule (see Fig. 5.1): suppose that a consistent orientation is imposed on each closed chain of edges bounding one face of the surface. Thus, two directional arrows are assigned to each edge. Then, this polyhedral surface is orientatable if, and only if, each edge has a pair of arrows pointing towards opposite directions.

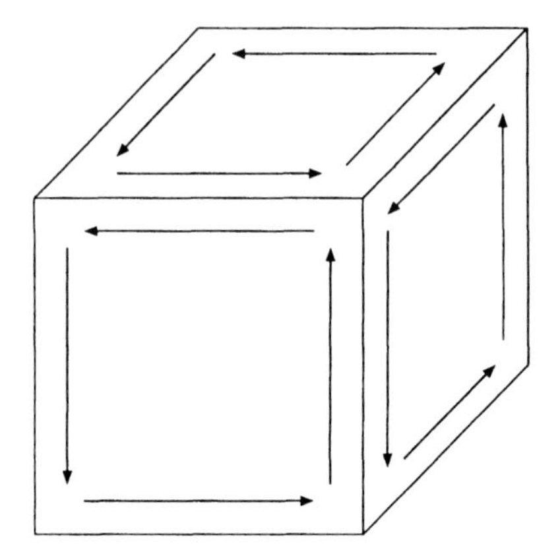

Figure 5.1: Checking for orientatability by means of the Möbius rule

In the following, we will use the term 'solid' in this sense. When appropriate, we will also carry over this term to 2D.

5.2.2 Formal Basic Demands of Modeling Schemes

3D Modeling schemes should meet the following basic demands (introduced by Requicha):

- **Validity**: It is required that for any given model there exists at least one real 3D object corresponding to this model. M. C. Escher's famous drawings of impossible objects provide examples for 'representations' not satisfying this criterion.
- **Unambiguity**: It is required that for any given model there exists at most one 3D object corresponding to this model.
- **Uniqueness**: It is required that for any given 3D object there exists exactly one model corresponding to this object (at least up to permuting 'substructures').
- **Representational completeness**: A representation scheme should, at least theoretically, be able to cope with a broad range of geometric interrogations.
- **Functional completeness**: It is required that all operations provided within the modeling system are applicable to all possible representations of objects within the used modeling scheme.
- **Large domain**: The domain of a modeling scheme is the set of objects of the real world that can be represented. Clearly, a poor domain drastically restricts the range of possible applications of a modeling scheme.

- **Conciseness**: The representation scheme should not contain redundant information (as this might cause troubles when updating a model).
- **Efficacy, user friendliness**: In order to minimize the computational efforts during interactions with the model the representation scheme should be well-suited for the intended applications.

One should observe that the last two principles usually contradict each other: Since the efficacy of algorithms operating on the models heavily depends on the representation scheme, it may be reasonable to keep different internal representations for one object. Clearly, such an approach introduces a lot of redundancy.

We would like to emphasize that a representation scheme that does not provide unambiguity or uniqueness needs not to be à priori inferior to others. For instance, wire-frame models, which are ambiguous and not unique as the reader will learn in Ch. 6.1.2, are entirely adequate for computing convex hulls of polyhedra.

5.3 Classification of CAD Systems by Dimensionality

CAD systems can be classified with respect to the dimensions involved as follows (see Fig. 5.2, [Cugini, 1988]):

1. dimension of the objects to be modeled,
2. dimension of the modeling space,
3. dimension of the modeling primitives.

Since in most applications 3D objects out of a 3D world have to be modeled, one usually distinguishes CAD systems with respect to the dimensionality of their modeling primitives. Thus, both 3D as well as 2D CAD systems enable the construction of 3D objects. In the case of a 3D system, the user inputs a spatial model of the object to be constructed. He may also use different cross-sectional and projective views in parallel. Furthermore, modifying a detail of the part in only one view simultaneously – theoretically, at least – affects the other views.

On the contrary, when working with a 2D system, one is restricted to using sectional and perspectives views. Unfortunately, modifying the object in one view has no immediate effects on the other views. Thus, the more the objects of the intended application are complicated, the more it is reasonable to use a 3D system.

However, in order to combine the main advantages of 2D and 3D systems – 2D systems are easy to use, 3D systems enable less tedious constructions of 3D objects – the idea of '$2\frac{1}{2}$D' has been introduced (see Fig. 5.3). Unfortunately, the term $2\frac{1}{2}$D has no precise definition; it merely indicates that not all aspects of 3D geometry are fully considered. Usually, geometric entities that can be specified as arrangements of parallel 2D layers are called $2\frac{1}{2}$D objects. Thus, by means of $2\frac{1}{2}$D systems one only is able to generate terrace-shaped 3D objects.

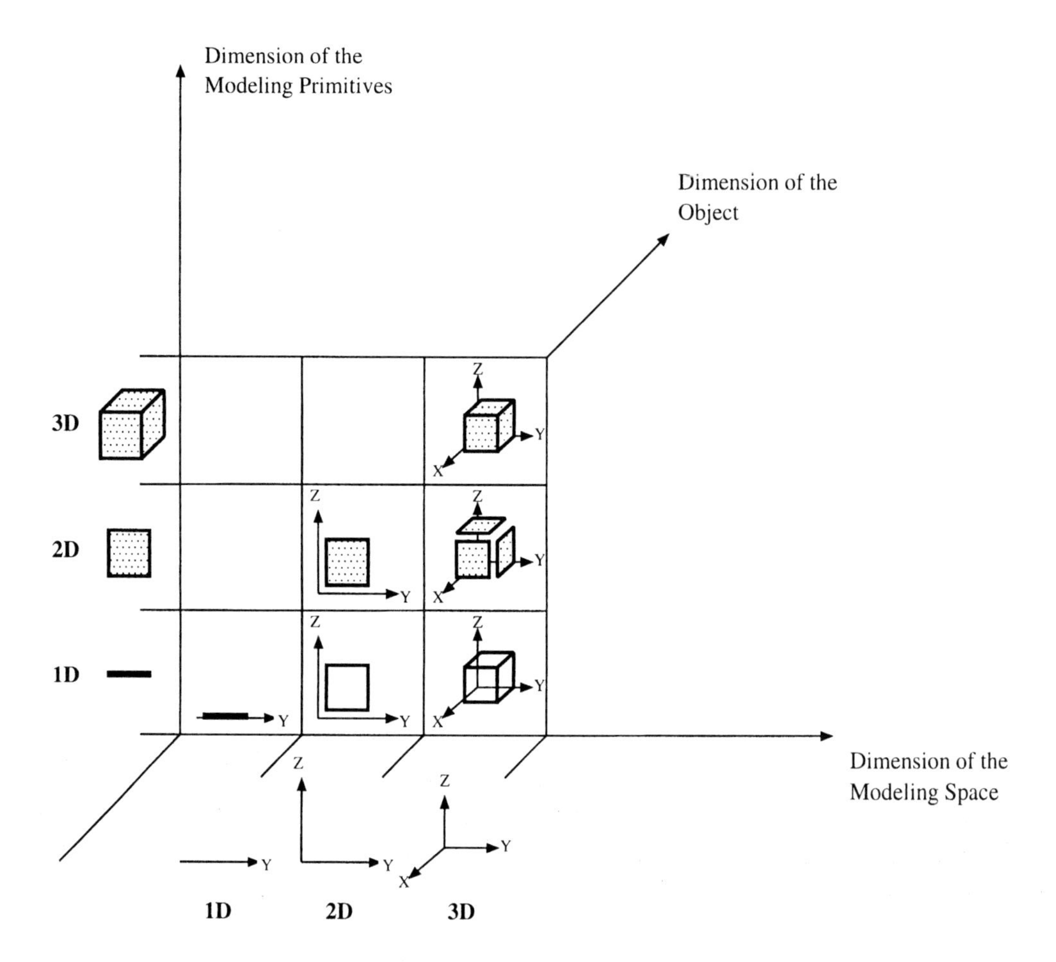

Figure 5.2: Different aspects of dimensions in CAD

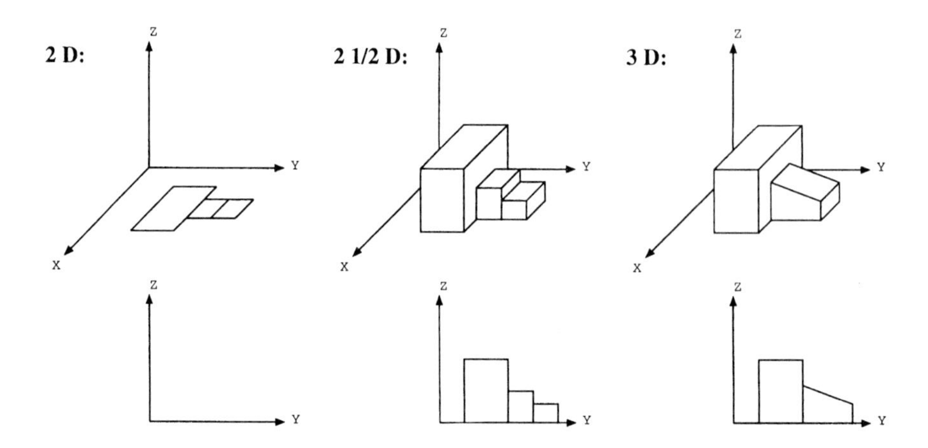

Figure 5.3: $2\frac{1}{2}$D systems constitute a hybrid of 2D and 3D systems

Sometimes, objects of revolution – i. e. objects that have a rotational symmetry – are considered to be $2\frac{1}{2}$D, too. Since in both of these two cases, parallel layers and rotational symmetry, most geometric aspects can be much more easily expressed than in the case of arbitrary 3D objects, $2\frac{1}{2}$D systems are widely used in spite of their modeling limitations.

5.4 Declarative versus Procedural Representations

There exist two conceptually quite different approaches to the problem of representing objects: declarative and procedural representations [Rooney, 1987b]. The difference between them is perhaps best illustrated by comparing the description of the state of the object to the process needed to obtain the object. For instance, a wire-frame model of a tetrahedron can be represented as a set of four vertices and six edges (declarative representation), or as an ordered sequence[1] of six directed edges (procedural description).

5.4.1 Declarative Representation

In more detail, this means that a declarative representation is usually built up from some primitive declarative forms and attributes, such as

- vertices, edges, and faces,
- points, lines, surfaces, and solids,
- position and orientation,
- equations and inequalities, etc.

Most of these relations are expressed using an algebraic, analytical formulation.

5.4.2 Procedural Representation

A procedural representation is usually built up from some primitive procedural forms and attributes, such as

- paths and cycles,
- pointers and records,
- translation and rotation,
- algorithms and procedures, etc.

Most of these relations are expressed using an iterative / recursive, synthetic formulation.

[1]The reader should observe that in general it is not possible to trace out a polyhedron unicoursally (i. e. in one sequence of edges).

5.5 Classification of Geometric Models by Their Representation Schemes

5.5.1 2D Representations

In 2D, one usually distinguishes between line-based models (i. e. 2D wire-frames) and area-based models.

5.5.2 3D Representations

Existing 3D representation schemes for complete representation of solids have been divided into the following six categories [Requicha and Voelcker, 1979], [Requicha, 1980]:

1. **Primitive instancing**: Families of objects are defined parametrically. A shape type and a limited set of parameter values specify an object.
2. **Spatial enumeration**: An object is represented by a list of the cubic spatial cells which it occupies.
3. **Cell decomposition**: A generalized form of spatial enumeration in which the disjoint cells are not necessarily cubic or even identical.
4. **Sweep representation**: A solid is defined as the volume swept by a 2D or 3D shape as it is translated along a curve.
5. **Boundary representation (B-Rep)**: Objects are represented by their enclosing surfaces (planes, quadric surfaces, free-form patches, etc.).
6. **Constructive solid geometry (CSG)**: Objects are represented as collections of primitive solids (cubes, cylinders, etc.). They are connected via the Boolean operations.

Additional to these schemes, the following incomplete methods are used:

1. **Wire-frame models**: An object is modeled by storing a collection of curve segments which represent the object's edges.
2. **Surface modeling**: Similar to B-Rep, objects are modeled by storing their surfaces but these models provide less topological information.

Besides these classifications, other schemes emphasizing the functional capabilities of CAD systems have also been proposed: see, for instance, the reference model of the German association of engineers [Abeln, 1989].

In the next chapter we will analyze the different representation schemes of geometric models in more detail.

Chapter 6

Geometric Models

This chapter gives a detailed survey on different geometric models in 2D and 3D. Mathematical problems of solid modeling are discussed and their solutions are pointed out for the different models. The focus is laid on volume modeling and does not consider surfaces and surface patches.

6.1 Wire-Frame Models

6.1.1 Wire-Frames in 2D

Wire-frame modelers are relics of the so-called drafting era of CAD. Usually, these systems use entities of drafting such as lines, circles, arcs, conics, splines, etc. for representing edges of the objects to be modeled. For a more precise definition of wire-frame models, the reader is referenced to [Tilove, 1981].

Unfortunately, these systems are much more calligraphic than modeling ones: in fact, they model the document rather than the object described in the document. There is no clear distinction between the (drawn) object and its representation. And even worse, it is all too easy to construct 2D wire-frame models that cannot be realized, i. e. for which there are no corresponding physical objects (see Fig. 6.1, [Goldman, 1987]).

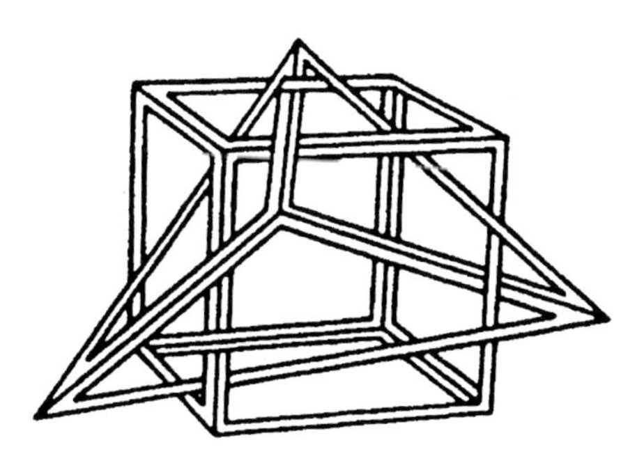

Figure 6.1: Unrealizable 2D wire-frame

Since in 2D systems there is no way to ensure the realizability of wire-frame objects, nonsense design readily will be passed on to manufacturing.

6.1.2 Wire-Frames in 3D

True 3D wire-frame models (see [Rooney, 1987a]) avoid some of these optical illusions. For instance, since the data is now really 3D, the situation depicted in Fig. 6.1 cannot occur because it is not possible for the extreme points of one line to be in front of a second line while the first line itself passes behind the second line. Nevertheless, it is possible to design ludicrous parts even when using a 3D wire-frame modeler (see Fig. 6.2, [Goldman, 1987]). Since each edge has to be input separately, the design is very tedious and prone to errors: the user must supply a lot of low-level data even for modeling simple objects such as a box! Thus, wire-frames tend to be incomplete because one or several edges are missing.

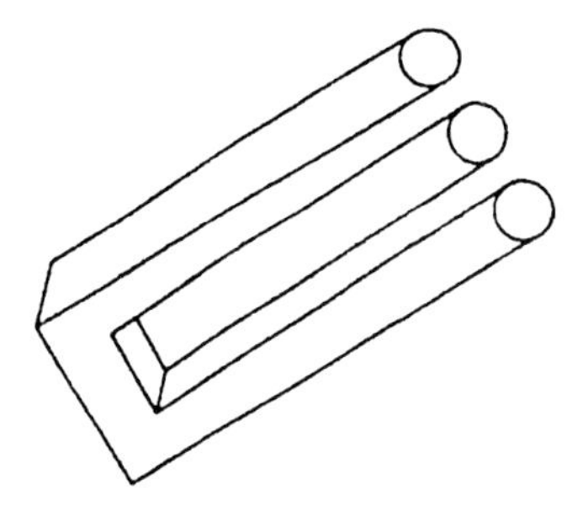

Figure 6.2: Impossible 3D wire-frame

Moreover, even if the wire-frame model is realizable it is not guaranteed that it is unambiguous (see Fig. 6.3, [Goldman, 1987]). A wire-frame, unless the object is a convex polyhedron, may correspond to more than one solid object and there are no standards for clearly discriminating between the different possible realizations. One should observe that in a production process ambiguity usually is worse than nonsense because the wrong part may be manufactured on the shop floor before the error is detected! An interesting approach to generating all possible solids that may be represented by a valid wire-frame is described in [Markowsky and Wesley, 1980]. How to create solids out of wire-frames has been illustrated in [Hanrahan, 1982].

Objects containing a lot of curved segments reveal another problem: silhouettes (i. e. profile lines; see Fig. 6.4) depend on the viewpoint. They are no physical edges and should therefore not be included into the model of an object: otherwise, after performing a rotation, their location has to be updated! But omitting these lines usually makes it difficult to imagine complex 3D scenes.

Furthermore, there are some other serious disadvantages:

- There is no information on the surfaces of the object. As an immediate consequence, there is no chance to determine the volume, to compute normal vectors, or to produce data for NC machining.
- Dividing a wire-frame into two halves does not work (see Fig. 6.5). More general,

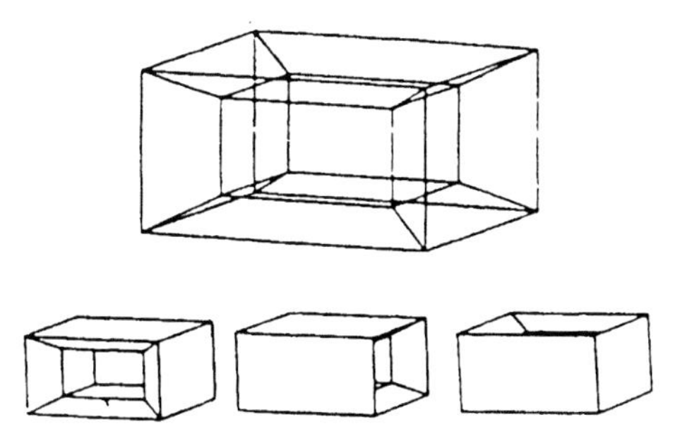

Figure 6.3: Wire-frame models are ambiguous

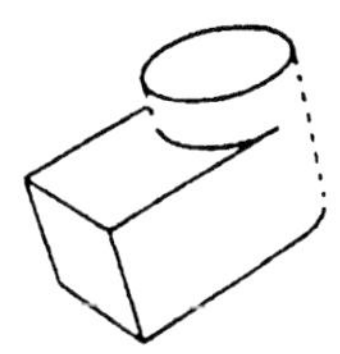

Figure 6.4: Profile lines (dashed) are viewpoint-dependent artifacts

union and intersection of two wire-frames cannot be computed and profiles cannot be obtained.

- Since wire-frames do not provide surface information, hidden-line elimination cannot be carried out. Thus, visualizing wire-frames is rather difficult.

Summarizing, wire-frame models have a lot of draw-backs: Some are unrealizable, many are ambiguous, most are incomplete, and all are tedious to construct and use.

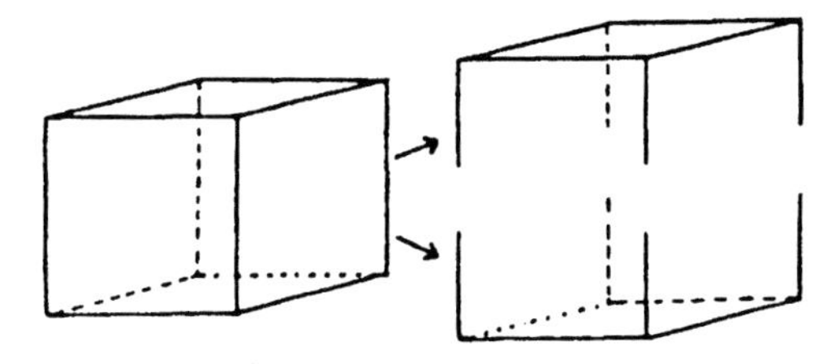

Figure 6.5: Dividing a wire-frame model into two halves does not work

6.2 Primitive Instancing

Primitive instancing is based on using classes of similar objects. Each member of a class is distinguished from the other members of the same class by a few parameters. Such a family of objects is called a *generic primitive*, and the individual members of a class are called *primitive instances* (see Fig. 6.6). Techniques for interactive modeling of parameterized objects have been described in [Gardan, 1983].

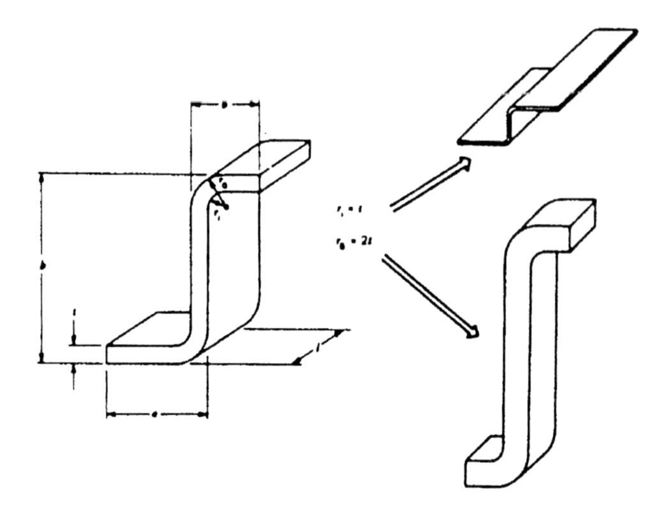

Figure 6.6: Generic primitives and two primitive instances

As a matter of principle, primitive instancing schemes are unambiguous, non-unique, concise and easy to use. Unfortunately, pure primitive instancing schemes do not offer means for creating new objects by combining the generic primitives (by means of set operations, for example). Thus, the domain of such a modeling system is restricted to the set of primitive instances.

Furthermore, it is difficult to ensure completeness because for each generic primitive a specific algorithm has to be designed. Writing algorithms for a large set of generic primitives may result in an immense amount of work. Hence, adding a new generic primitive may necessitate extensive development of mathematical tools and significant software modifications.

6.3 Sweep Representation

6.3.1 Simple Sweeping

A popular, unambiguous representation scheme is called (simple) sweeping. The main idea of a sweep representation is to represent a solid set of points as the Cartesian product of an area set and a trajectory set. As this simple approach can only produce extruded objects (see Fig. 6.7a, where dashed lines indicate the edges of the swept profile), the sweep technique is usually combined with some other representation method resulting in hybrid representation schemes. For example, it is common practice to add a restricted form of a union operator (the so-called gluing operator[1]). This gives the possibility to model objects as depicted in Fig. 6.7b. A halfspace representation for extrusions has been published in [Peterson, 1984].

[1]The gluing operator is a restriction of the union operator. It is only applicable to objects with non-intersecting interiors.

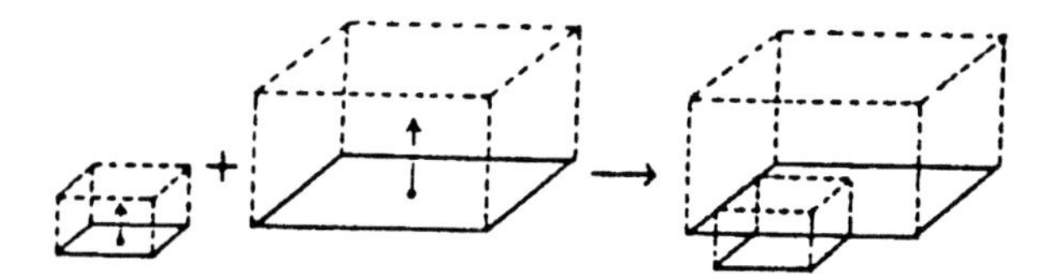

Figure 6.7: Pure and hybrid sweep representations

6.3.2 Generic Sweeping

A more powerful method of modeling is provided by generic sweeping: Suppose that a planar face – a cross-section – and a main-flow line – a so-called spine – is given. Then, a new object can be modeled by sweeping the face along the spine, i. e. by moving a fixed point of the planar face along the spine such that the spine remains perpendicular[2] to the plane in which the face lies (see Fig. 6.8).

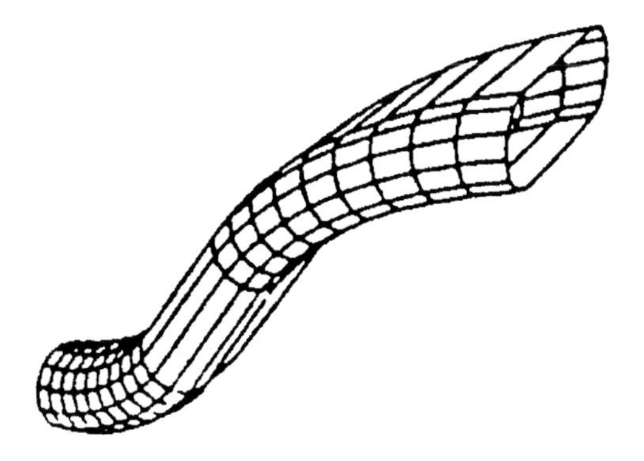

Figure 6.8: Generic sweeping

By means of generic sweeping, objects like pipes and ducts can easily be modeled. Controlling the twist may be achieved by using local coordinate systems.

Care has to be taken that the orthonormal plane to the spine is defined at each point of the spine. Furthermore, one has to pay attention that portions of the surface generated by a sweep are eliminated if there are self-intersections. Otherwise, one readily will run into serious troubles when manufacturing such a part. Self-intersections may occur at points of the spine that have a rather large curvature.

Furthermore, although sweeping is conceptually easy understood, one should observe that it is less understood seen from a theoretical point of view. For instance, it is difficult to avoid dangling edges and similar dimensionality problems (as depicted in Fig. 6.9, [Mortenson, 1985]). Few is known about mathematical conditions in order to ensure that the result of generic sweeping is a regular set. This is even more difficult if the swept profile is considered to be non-rigid, i. e. if it undergoes deformations (such as shrinkage) as it is swept through space.

[2]Some modeling systems support an inclination of the plane of the face with respect to the spine, too.

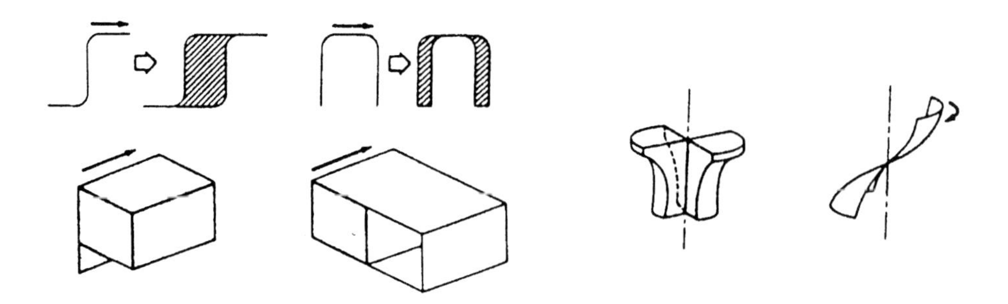

Figure 6.9: Sweeping may produce non-manifold objects

6.3.3 Generating Special Types of Surfaces by Means of Sweeping

Tabulated Cylinder

As depicted in Fig. 6.10a, a *tabulated cylinder* (*translational sweep*) $T(u,v)$ is generated by moving a profile $P(u)$ for a specified distance along a specified axis $A(v)$:

$$T(u,v) = P(u) + A(v).$$

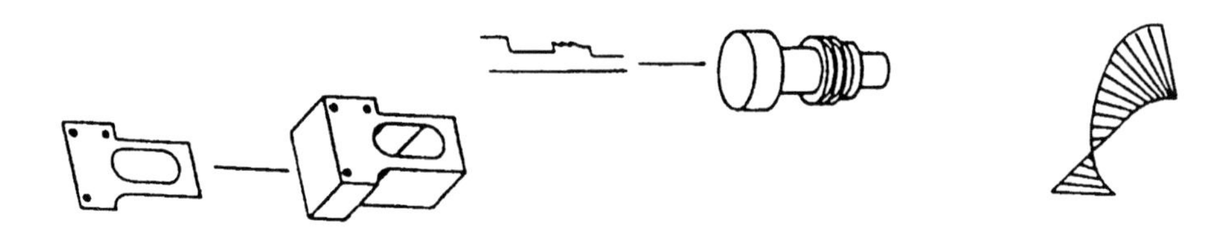

Figure 6.10: Tabulated cylinder, surface of revolution, and ruled surface

Surface of Revolution

A *surface of revolution* $R(u,\vartheta)$ is generated by rotating a profile $P(u)$ through a specified angle around a specified axis (*rotational sweep*; see Fig. 6.10b). Suppose that P is given by $P(u) = (r(u), 0, f(u))^T$ and that the z-axis is the rotation axis. Then,

$$R(u,\vartheta) = (r(u)\cos\vartheta, r(u)\sin\vartheta, f(u))^T.$$

Ruled Surface

A *ruled surface* $R(u,\lambda)$ is generated by a straight line the extreme points of which are continuously moving along two curves $C_1(u)$ and $C_2(u)$ (see Fig. 6.10c). Thus, the ruled surface is given by the following convex combination:

$$R(u,\lambda) = C_1(u)(1-\lambda) + C_2(u)\lambda.$$

The reader should observe that the shape of a ruled surface heavily depends on the parameterizations of C_1, C_2, not only on their shape!

6.4 Surface Modeling Schemes

Surface models, see [Pratt, 1987], have been introduced in order to overcome the deficiencies of wire-frame models. Two common types of surface modeling schemes are widely used: transfinite interpolation and discrete approximation. Before we are discussing these main types free-form surfaces, let us have a survey of surfaces incorporated in surface models.

6.4.1 Primitive Surfaces

As stated in [Encarnaç
ao and Schlechtendahl, 1983, Goldman, 1987], modeling systems usually use one or several of the following classes of surfaces:

- **Plane**: Although most industrial products have curved surfaces, solid modeling based on planar surfaces has a wide range of practical applications: architectural design and finite element analysis are usually based on polyhedral representations, for instance. The same principle of planar approximation may also be utilized for other applications such as hidden-line and hidden-surface removal. Although an accurate polyhedral approximation may produce quite a lot of planar patches, this is not that bad because the algorithms for dealing with these patches are simple and very efficient (at least, if compared to the corresponding algorithms for more complex surfaces).

  When using a polyhedral representation, there is no possibility for modeling common features such as holes or fillets. Furthermore, free-form design and complex blending is at least difficult.

- **Quadric**: The attempt to build solid modeling systems thereby keeping the mathematics as simple as possible led to the introduction of quadric surfaces. Quadric surfaces are defined by second-order polynomials in the three coordinates (such as spheres, cones, etc.).

  In [Sarraga, 1983], methods for analytically computing and representing the intersection of two quadrics are given. Thus, at least theoretically, it is possible to convert from a CSG tree to B-Rep when using quadrics. Furthermore, since quadrics are second-order surfaces, ray tracing of quadrics requires solving only a second-order equation.

  Usually, only a special subset of quadrics is implemented because not all quadric surfaces are used in design. For example, hyperbolic paraboloids are hardly ever used. An application of quadric surface models is illustrated in [Fuh *et al.*, 1985].

- **Torus**: Since quadrics are only second order surfaces they cannot turn corners. Thus, the torus – which is not a quadric surface – is introduced for bending around corners. Furthermore, tori are used for blending and filleting (see Fig. 6.11). Ray tracing of tori requires the solution of an equation of degree four. Hence, tori still can rather easily be rendered.

  Unfortunately, intersecting two tori yields an algebraic curve of degree sixteen. Thus, unlike quadrics, it is no longer practical to compute and store an exact analytical representation of the intersection of two tori. But approximating the intersection curve by some numerical techniques makes it nearly impossible to consistently convert between CSG and B-Rep!

Figure 6.11: Part of a torus as a result of blending

Quadrics and tori together usually are sufficient for modeling most rigid mechanical parts but they are not capable of modeling complex blends and free-form surfaces.

- **Superquadric**: Roughly, superquadrics are quadrics that are deformed along their main axes (see [Zarrugh, 1985]). These additional degrees of freedom make superquadrics useful for approximating various shapes (see Fig. 6.12).

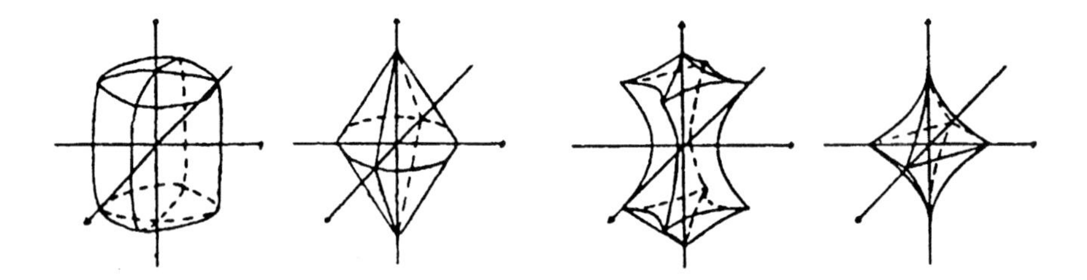

Figure 6.12: The versatility of superellipsoids

Superellipsoids are given by the implicit equation

$$(|\frac{x}{a}|^{2/e_2} + |\frac{y}{b}|^{2/e_2})^{e_2/e_1} + |\frac{z}{c}|^{2/e_1} = 1,$$

where $a, b, c \in \mathbf{R}^+$ denote the conventional coefficients of quadrics and $e_1, e_2 \in \mathbf{R}^+$ are two shape parameters. The parameterized representation of superellipsoids is given by

$$(x, y, z) = (a \cos^{e_1} \alpha \cos^{e_2} \beta, b \cos^{e_1} \alpha \sin^{e_2} \beta, c \sin^{e_1} \alpha),$$

where $\alpha \in [-\frac{\pi}{2}, \frac{\pi}{2}], \beta \in [0, 2\pi]$. The smaller e_1, e_2 are, the more the shape of the superellipsoid is box-like (with faces nearly parallel to the coordinate planes). Obviously, for $e_1 = e_2 = 1$ these equations represent an ordinary ellipsoid. For $e_1 = e_2 = 2$, a polyhedron is generated. Furthermore, one should remark that the superellipsoid is convex for $e_1 \leq 2 \wedge e_2 \leq 2$.

Due to their capability of representing portions of various shapes, superquadrics would constitute means for guaranteeing high flexibility. Unfortunately, handling a superquadric is no that easy. For instance, for $e_1, e_2 \neq 2^{-k} (k \in \mathbf{N})$, the implicit equation is no longer algebraic.

- **Cyclid**: The concept of Dupin cyclids (see, e. g., [Nutbourne and Martin, 1988]), applied to geometric modeling, permits a unified representation of the quadrics, of a generalized squashed torus, and of their offsets (i. e. the offset of a cyclic surface is a cyclic surface). Furthermore, difficult blends are supported. Cyclids are a special case of rational bi-quadratic surfaces. Unfortunately, while the closure of cyclids under offsetting is a nice property, certain aspects of cyclic surfaces seem to be too restrictive for free-form modeling. For instance, the four corners of a patch lie on a cyclic quadrilateral (i. e. a quadrilateral that can be inscribed in a circle). Furthermore, all main lines of curvature are circles. On the other hand, this simplicity gives the possibility to obtain some useful analytical results.

- **Free-form surface**: Free-form modeling includes (non-uniform) rational B-splines (so-called NURBS, see [Tiller, 1983]), (rational) beta-splines (see [Barsky and Beatty, 1983]), and other spline-based modeling techniques. [Ocken *et al.*, 1987] gives a fine introduction to the usage of rational parameterizations of basis primitives, a topic which is steadily gaining in importance. Using a rational representation is motivated by the fact that it gives the possibility to store planes, spheres, general quadrics, and free-form elements within one unique representation scheme.

6.4.2 Transfinite Interpolation

The goal of transfinite interpolation is to construct a surface that passes through a given site of curves (i. e. interpolates a transfinite number of points). The curves themselves may represent cross-sections or contour lines.

Two basic methods have been proposed for transfinite interpolation: one is local, the other global. Coons [Coons, 1967] developed a local four-sided method, the so-called Coons patch, which is both defined and interpolates to its four boundary curves. By connecting many Coons patches, one is able to efficiently build up quite large surfaces.

A well-known global method has been developed by Gordon (see [Gordon, 1969]). Using cubic cardinal splines to interpolate any arbitrary rectangular mesh of curves, the so-called Gordon surface is constructed.

Both techniques are widely used for fitting a surface on a wire-frame skeleton. Since the generation of the surface is controlled by means of curves, both methods are dominated by wire-frame techniques.

6.4.3 Discrete Approximation

In the case of discrete approximation the user specifies only a relatively small amount of scattered data, the so-called control points. The modeling system then generates a surface to approximate the shape described by these points. In general, it is not requested that the surface has to interpolate the control points. By changing the location of the control points the shape of the surface is modified.

As in the case of transfinite interpolation, discrete approximation also is performed by applying global and lócal methods. A popular global method is due to Bézier [Bézier, 1968]. Using Bernstein polynomials a single polynomial surface is constructed – a so-called Bézier patch. Although widely used, this method suffers from two main disadvantages: First, if there are a lot of control points then high-order polynomials have to be used. Furthermore, modifying a single control vertex may already cause a dramatic change of the overall shape of the surface. Thus, there is no local control of the surface shape.

A well-known local method has been introduced by Riesenfeld [Riesenfeld, 1973a], [Riesenfeld, 1973b]. Using B-splines, a piecemeal polynomial surface is constructed guaranteeing smoothness between adjacent patches. A lot of control points can be handled increasing the number of patches rather than increasing the degrees of the polynomials. Furthermore, since B-splines have local support, local control of the shape is enabled.

6.4.4 Applicability of Surface Modeling

Clearly, it is advantageous to take surface models instead of wire-frames. For instance, the impossible objects depicted in Fig. 6.1 and Fig. 6.2 cannot be constructed using real surfaces. And by incorporating surfaces in the computer model many wire-frame models are no longer ambiguous. Furthermore, free-form shapes such as ship hulls, car bodies, turbine blades, or aircraft wings can easily be represented using surface modeling systems.

Unfortunately, when dealing with surface modelers, the validity of the constructed models cannot be guaranteed. For example, the famous Klein bottle can be represented as a surface model although it cannot be manufactured on the shop floor (because it does not exist in 3D). Moreover, since primitive objects cannot be used, there is no chance to reduce the construction time in the case of repeated design of similar objects.

Even more serious is the fact that surface models are often incomplete (i. e. it is not ensured that the surfaces bound a true solid object). If an incomplete model is passed to an human expert he perhaps will detect the error and then time-consuming corrections have to be undertaken. If an incomplete model is passed to another computer program (such as programs for NC code generation) then erroneous data is usually computed. In the case of controlling an NC machine such errors may have fatal consequences (such as the destruction of the workpiece or of the cutting tool).

Furthermore, there usually is no chance to determine whether or not two surfaces are adjacent (by simply operating on the data base of a surface model). This lack of topological information makes it impossible to automatically perform tasks such as NC verification.

6.5 Boundary Representation

Boundary representation models (B-Reps) are familiar to most computer scientists due to their use in computer graphics. A solid is represented by segmenting its boundary into a finite number of bounded subsets – so-called faces or patches – and representing each face by some means (see Fig. 6.13). Furthermore, topological information is explicitly stored. Thus, the B-Rep representation yields a directed graph containing the object as root and the faces, edges, and vertices as nodes.

B-Rep schemes are potentially capable of covering large domains. They are unambiguous if the faces are represented unambiguously. This follows from mathematical theorems which

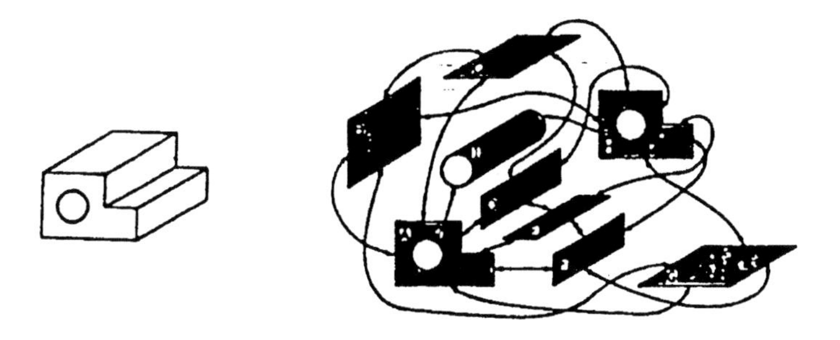

Figure 6.13: B-Rep of a simple workpiece

ensure that a regular set is defined unambiguously[3] by its boundary. In general, a B-Rep is not unique.

6.5.1 Ensuring the Validity of a B-Rep Model

6.5.1.1 Conditions for Validity

The validity of B-Rep models is quite difficult to ensure. When trying to verify the validity of a model, one usually proceeds bottom-up in the representation tree by investigating conditions for vertex nodes, for edge subgraphs (i. e. edge nodes plus corresponding vertex nodes) representing curve segments, and so on.

In [Requicha, 1980], validity conditions are stated for triangulation schemes. The following *topological conditions* are distinguished:

- Each face must have exactly three edges.
- Each edge must have exactly two vertices.
- Each edge must belong to an even number of faces.
- Each vertex of a face must belong exactly to two of the face's edges.

However, topological conditions are not sufficient for ensuring the validity of an object; the following *geometric conditions* must be satisfied, too:

- Each triple of vertex coordinates must represent a distinct point of $\mathbf{E}^3$.
- Edges must be disjoint or intersect at a common vertex.
- Faces must be disjoint or intersect at a common edge or vertex.

One should observe that these three conditions require face-face testing which may be computationally expensive if the validity conditions are generalized to more complex faces.

[3]For general, non-regular sets of $\mathbf{E}^3$ this is not true.

6.5.1.2 The Concept of Euler Operators

The concept of Euler operators (see [Eastman and Weiler, 1979]) is based on a well-known theorem due to Euler (which is known as *Euler's formula* for polyhedra): Let V, E, F denote the number of vertices, edges, and faces of a simple polyhedron[4]. Then,

$$V - E + F = 2$$

provides a simple relation between these numbers. This formula has been generalized to n-dimensional space by Poincaré [Mäntylä and Sulonen, 1982].

Fortunately, Euler's formula is not restricted to polyhedra but it can also be applied to any closed surface on which a proper net – i. e. a net consisting of patches, curve segments and vertices – can be constructed. A proper net has to meet the following conditions:

1. The faces enclose a singly connected object[5].
2. All faces are single connected and bounded by a single ring of border elements. Hence, all faces have to be topological discs.
3. Each edge is common to exactly two faces and is terminated by one vertex at each end.
4. Each vertex is common to at least three edges.

Dealing with a bounded portion of space consisting of C polyhedral cells, Euler's formula can be modified yielding the following relation between vertices, edges, faces, and cells:

$$V - E + F - C = 1.$$

Deleting conditions 1) and 2) gives the possibility to handle multiply-connected objects. As stated in [Braid *et al.*, 1978], again by generalizing Euler's formula one gets the relation

$$V - E + F - H = 2(B - P),$$

where H,B, and P denote the number of holes in faces, the multiplicity of the object (i. e. number of separate disjoint bodies), and the genus of the object (i. e. the number of holes through the entire object), and where V,E, and F retain their usual meaning. Of course, all variables have to be non-negative.

Fig. 6.14 illustrates the above stated formulas. The reader should observe that the conventional Euler formula must not be applied to the wire-frame depicted in Fig. 6.14a (although it yields a valid result)!

In order to ensure the topological validity of a B-Rep model under construction (or under modification), it seems to be desirable to satisfy these formulas. More precisely, this means that the application of a construction step should result in a transition between two stages each of which is satisfying (one of) these formulas. A set of operators that meets this demand came to be known as *Euler operators*. For instance, an operator that splits an edge into two parts should know that a new vertex has to be inserted in the interior of the original edge. For a nice example of a tetrahedron being constructed by means of Euler operators, reference is given to [Mortenson, 1985].

[4]A polyhedron is called simple if it can be continuously deformed into a sphere.

[5]Informally, an object is singly connected if it has no holes through it.

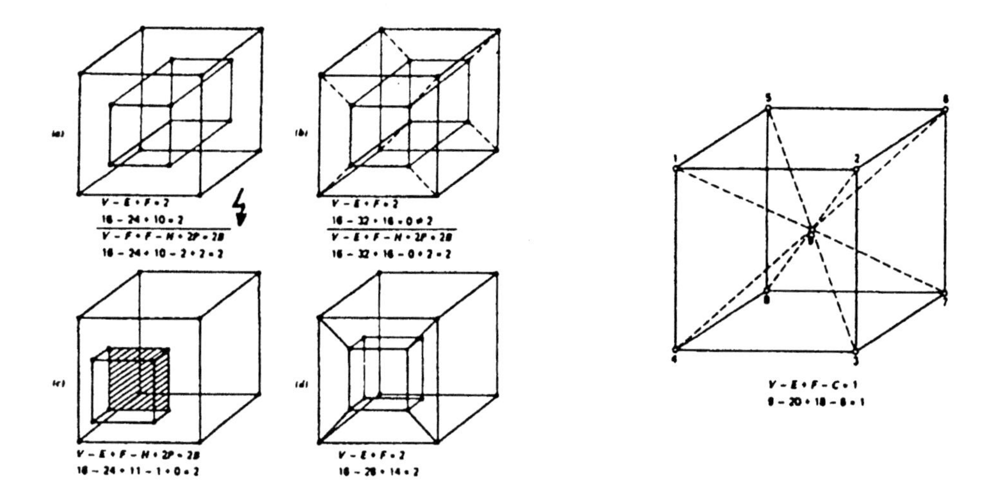

Figure 6.14: Applying the (modified) Euler formula

6.5.2 Polyhedral Models

Approximation by planar facets is a special case of B-Rep. In this case, the modeling system only deals with polyhedra (with many planar facets bounded by straight lines). Hence, the kernel of such a modeler can be relatively simple and can take advantage of hardware implementation and parallel processing. Using special hardware may result in a considerable speed-up of the execution time of application programs. For instance, this modeling scheme enables the user to take benefit of standard features of computer graphics systems (such as polygon filling, hidden line / hidden surface removal (HL / HSR), shading). As emphasized in [Ying and Zhou, 1988], set theoretical operations on triangle-based models can readily be executed by intersecting its simple facets (i. e. triangles) rather than complex surfaces. A more general approach to intersecting polyhedral models has been presented in [Thibault and Naylor, 1987].

Furthermore, there is no problem with introducing a new family of shapes or surfaces. Only one new approximation algorithm is needed in order to generate the facets (as compared to the case of surface models and traditional B-Reps where n new intersection algorithms have to be implemented for determining the intersections of the new family of surfaces with any of the n already existing ones).

The major drawback of this scheme is the problem of guaranteeing sufficient accuracy and precision. Clearly, there is a trade-off between the gain in simplicity of the algorithms and the increasing verbosity of the modeling data. Guaranteeing precision becomes even more difficult if a lot of operations and transformations have to be performed on the model.

6.5.3 Applicability of B-Rep Modeling

Boundary representations are verbose. The large amount of data needed makes it difficult to model without computer assistance. For this reason and since correct (i. e. formally valid)

B-Reps are difficult to construct – because they must satisfy sophisticated conditions, as described above – it is common practice to construct this representation from other ones using conversion algorithms.

The main advantage of a B-Rep is the ready availability of full information on geometric and topological details of the modeled object. Information on faces, edges and on the relations between them is extremely important for generating graphic displays (such as line drawings) and for manufacturing purposes. Clearly, a machining algorithm needs detailed information on the surfaces of the workpiece to be manufactured.

B-Reps tend to be numerically unstable: Since the representation is a composite of real numbers (geometry) and integers (topology), it is subject to numerical errors. The basic difficulty arises from the need to make local numerical decisions (such as whether a vertex is located above, below, or on a face) without being able to check for global consistency. Thus, the model of an object may suffer from a global inconsistency although it is locally consistent. Furthermore, the validity of B-Rep models usually is dependent on some tolerance bound because numerical tests can only be performed with respect to a tolerance.

6.6 Constructive Solid Geometry

6.6.1 Representing Objects as CSG Trees

Constructive solid geometry (CSG; [Boyse and Gilchrist, 1981]) stands for a family of representation schemes where rigid solids are constructed performing Boolean combinations (see Fig. 6.15) and transformations of some basic solid components. These so-called *convenience primitives* usually include boxes, cylinders, spheres, spherical caps, truncated cones, fillets, pipes, tori, etc. Some CSG systems also allow the usage of user-defined primitives.

Figure 6.15: Boolean operations on two solids

CSG representations are ordered binary trees[6] (so-called *CSG trees*, see Fig. 6.16). Intermediate nodes represent operators, which may be either rigid motions or Boolean combinations. Terminal nodes are either convenience or generic primitives, or transformation leaves containing the defining arguments of rigid motions.

[6]It is common practice to share subtrees. Hence, the CSG representation usually is rather a graph than a tree.

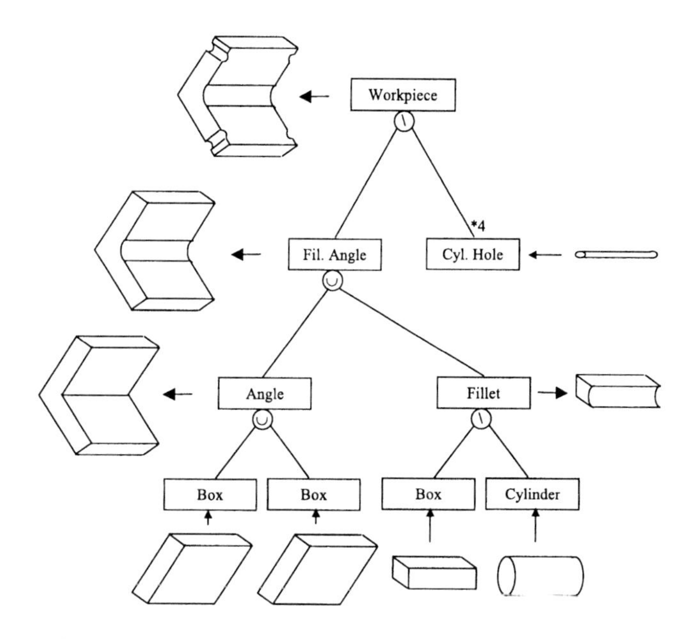

Figure 6.16: A CSG object represented as a binary tree

6.6.2 Ensuring the Validity of a CSG Model

Let the regularized intersection of two sets be the closure of the interior of their conventional set-theoretical intersections. Similarly, the other regularized set-theoretical operators are defined. Regularized rather than conventional operators are used to ensure that the result of a Boolean combination of two regular sets is a regular set (see [Requicha and Voelcker, 1977]). As depicted in Fig. 6.17, this is not true for conventional set-theoretical operations. By convention, regularized Boolean operators are denoted by adjoining a star to the symbol of the conventional operator. For instance, $\bigcap^*$ denotes regularized set intersection.

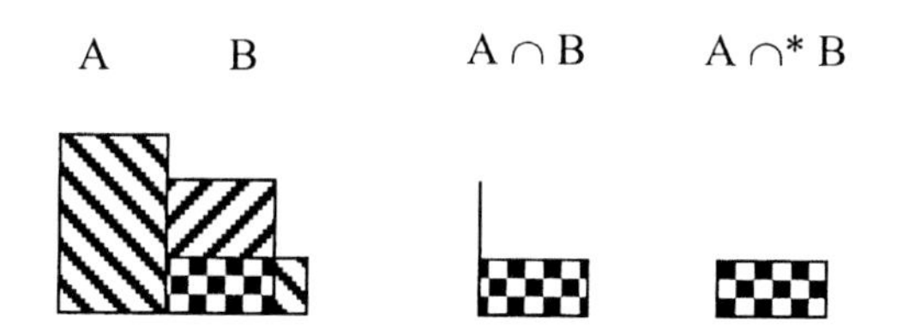

Figure 6.17: $A \cap B$ yields a dangling edge (dashed)

Provided that the primitives are regular sets and that the regularized complement is not used[7], it is guaranteed that any CSG tree represents a valid physical object (if all primitives are bounded[8]). Thus, the validity of CSG trees can be ensured by guaranteeing the validity of each of its primitives stored in the leaves of the tree; i. e. validity may essentially be ensured

[7]Using the regularized complement would destroy boundedness.

[8]CSG trees in schemes based on unbounded primitives may represent unbounded sets and may therefore be invalid. Unfortunately, boundedness of a set is difficult to verify. The only known way is to re-represent the object by means of its boundary and to test for boundedness (which may be computationally expensive).

at the syntactical level. Similarly, the validity of CSG trees with non-instantiated parameters – representing so-called macro objects or generic objects – is easily ensured.

Anyway, the reader should observe that it may be pretty difficult to perform Boolean operations on complex primitives within a reasonable amount of time [Hook and Tiller, 1989].

6.6.3 Restricted CSG Schemes

CSG-like schemes whose operators are not applicable to all pairs of objects in the domain are perhaps being used more extensively than proper CSG schemes. Usually, restrictions are imposed in order to facilitate the boundary evaluation, i. e. the computation of an object's boundary (see [Bhat and Aziz, 1988]).

Clearly, restricting a CSG scheme – for instance, by using a glue operator rather than a union operator – shrinks the domain of objects that can be modeled. This may result in an inadequacy for specific operations: for instance, checking for collision of two objects (i. e. for interference) does not work without a general intersection operator. However, there are more important drawbacks:

- The result of applying an operator to objects may in general not be used as input for other operators (see [Tilove and Requicha, 1980]). More serious, it is necessary to check whether an operator is applicable or not. In the case of using gluing, this means that one has to test whether the interiors of two objects to be combined do not intersect.
- It is important to remark that it is much more difficult to ensure the validity of objects if restricted operators are used. According to [Requicha, 1980], the validity properties of restricted CSG schemes are comparable to those of cell decompositions.

6.6.4 Applicability of CSG Modeling

Performing Boolean operations, a nearly unlimited variety of parts can be constructed from a rather small set of primitives. The representation is very concise, especially if bounded primitives are used[9]. Usually, a boundary representation of an object needs much more storage than the corresponding CSG model. CSG schemes are always unambiguous but usually not unique (see Fig. 6.18).

The Boolean operations mirror the manufacturing process. For example, subtracting a cylinder is equivalent to drilling a hole, welding two parts is equivalent to gluing. Hence, a CSG tree not only contains a mathematical description of a workpiece, but also information for establishing a manufacturing plan – at least theoretically.

Producing line drawings is rather difficult when using a CSG representation. Similar, it is difficult to support interactive actions such as pick an edge. However, it is easier to pick a face and to generate shaded displays.

[9]Handling unbounded primitives often requires more and lower-level primitives.

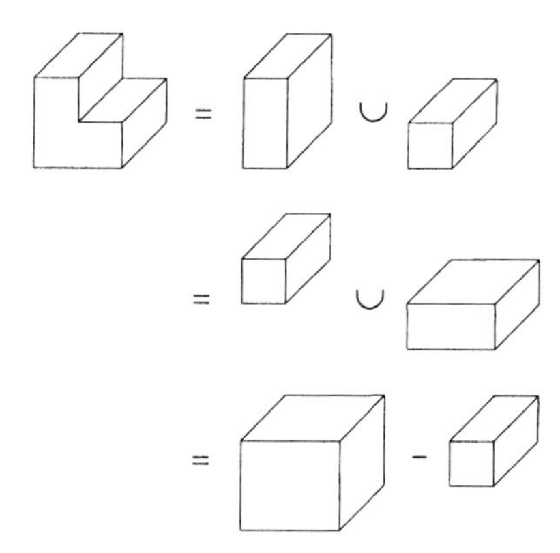

Figure 6.18: A CSG definition is usually not unique

6.6.5 Comparing CSG Modeling to B-Rep Modeling

Both CSG and B-Rep are commonly used modeling techniques. They can be compared as follows:

- CSG is concise whereas B-Reps are verbose. B-Reps produce a large amount of data (which is about an order of magnitude larger than the corresponding CSG representation).
- CSG models are guaranteed to be valid, largely independent from the number representation of the computer. Mapped into a character string and transmitted to another machine, a CSG tree may not represent the required object, but at least it is guaranteed to be valid. This usually is not true for B-Reps. Most B-Rep modeling systems still operate in a (computer aided) user-be-aware mode.
- B-Reps tend to be numerically unstable: Since the representation is a composite of real numbers (geometry) and integers (topology), it is subject to numerical errors.
- Algorithms for obtaining a B-Rep from other representation techniques are known. Converting to CSG is usually difficult or impossible.
- B-Reps are widely accepted in the computer graphics community. They are more closely related to drafting than is CSG. For instance, it is rather easy to generate wire-frame drawings from a B-Rep.
- The applicability of CSG and B-Rep seems to be complementary. Hence, dual CSG / B-Rep systems have been proposed.
- Using a B-Rep, the restrictions of a CSG domain are surmounted. For instance, the available technology for dealing with sculptured surfaces is much easier incorporated into a B-Rep based system rather than in a system based on CSG.

6.7 Representation by Means of Approximation

6.7.1 Motivation for Using Spatial Enumeration Techniques

CSG modeling, wire frame modeling and other similar techniques presented in the previous sections have several disadvantages that make their use for the modeling of dynamic objects, i. e., objects that change their shape with time, rather problematic. The major disadvantages are:

- Generally, it is not the status quo of the object that is modeled but the original object together with the history of the dynamic changes of size or shape. This leads to – generally – a rather exorbitant memory consumption of the computer internal model and makes the handling time dependent on the number of changes that have been performed on the model. A CSG model of a workpiece being machined, for instance, has a memory consumption that is of the order $O(n^4)$, where n is the number of tool movements [Drysdale *et al.*, 1989].
- CSG and similar models generally do not constitute a canonical geometric model. This means that the generation of a model from the real object is not a unique process, i. e. the very same object can be represented by rather different computer internal models. Furthermore, model comparison is generally not possible. This means that for most practical applications it is too time-consuming to check by computer whether two computer internal models represent the same real object. This fact is a great disadvantage for many applications. For example, when simulating an NC machining process, one wants to check whether the generated model of the machined workpiece is – up to a specified exactness factor – the same as the desired one. For details on canonical geometric models, see [Mayr and Stifter, 1989].

These and other disadvantages of CSG style modeling have increased the popularity of spatial enumeration techniques considerably (see [Carlbom *et al.*, 1985, Hersh, 1988]).

The principle idea of *spatial enumeration techniques* is the following: The modeling space is partitioned into volume elements that may either be of same size and regularly distributed or adapted to some characteristic values (e. g. coordinates of the object to be modeled).

Now this type of modeling offers the following advantages:

- It generally supplies a canonical geometric model.
- The current status of a dynamic object is stored, which allows efficient operations and easy coupling of dynamic process modeling.
- Generally, no memory explosion takes place.
- Many operations can easily be reduced to very simple operations on the primitives. Therefore, it is often possible to parallelize the operations.

However, there is one major disadvantage that has to be handled:

- The representation of the objects is not exact but – generally – some sort of approximation by kinds of rastering techniques.

This problem is not so critical as it seems to be at first glance. The reason is that when modeling objects in a computer, approximations have to be done at several phases (see Fig. 6.19). However, research has been done to improve the different approaches to get a better representation of the objects or to combine several ways of modeling (*hybrid modeling*) to have the best suited computer internal model for any of the intended operations.

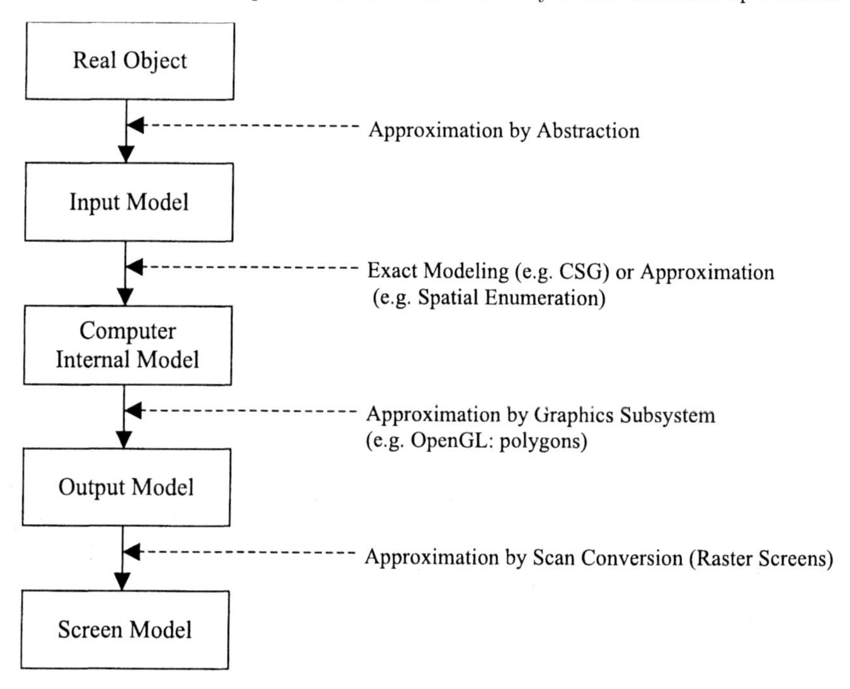

Figure 6.19: Approximation steps during the modeling of a real object

In the following sections, we will present a classification scheme for spatial enumeration techniques [Heinzelreiter and Mayr, 1990]. According to this scheme, the major modeling methods using spatial enumeration are then classified. After that, the advantages and disadvantages of these methods are discussed by investigating their applicability for modeling a dynamic object.

6.7.2 A Classification Scheme for Spatial Enumeration

Primarily, one can distinguish two classification criteria for spatial enumeration techniques, namely

1. their *global structure* (dealing with object hierarchy and data structures), and
2. their *local structure* (dealing with regularity).

The first criterion is called *global* since it influences the whole modeling process, starting from analyzing the object to specifying a suitable data structure, etc. The second criterion is called *local* since it deals more with the local way of representing (and approximating) an object.

6.7.2.1 The Global Structure

This structure mainly consists of three levels of hierarchy for a computer internal model of the real object (Fig. 6.20):

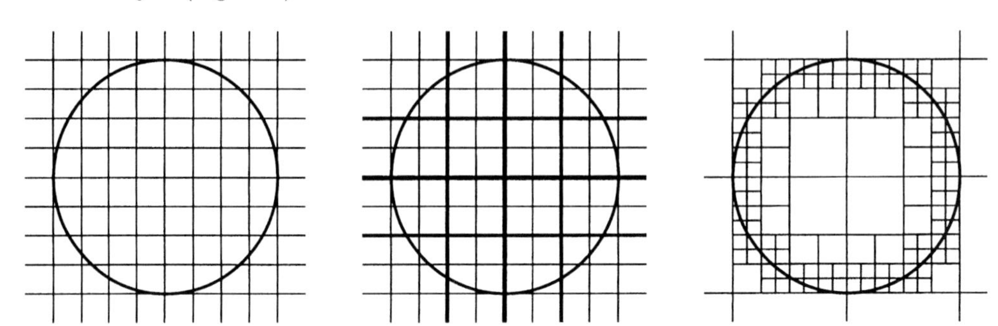

Figure 6.20: Different hierarchy levels for modeling an object

1. **Non-hierarchical modeling structure:** In this type of modeling, every single volume element is on the same level as all the others. The information is restricted to its single volume area, the access can (and has to) be done directly and for each volume element.

   Advantages are simple operations (operations for the whole model are just the repeated operations on a single element), direct access to the elements, and rather easy visualization. The main disadvantage is the huge memory consumption. The typical data structure is the matrix.

2. **Redundant-hierarchical modeling structure:** For this structure, the non-hierarchical modeling structure is extended by a hierarchical concept that is put on top of the structure. This means that for a certain number of volume elements at one level, there is one parent element on the next higher level that contains already information about the child volume elements. However, the volume elements at the lower level are stored even when all of them are homogeneous (i. e. they contain the same information). In this way, the advantages of redundancy (e. g. visualization) and hierarchy (e. g. Boolean operations) are combined.

   Advantages are fast access, simple operations, and easy visualization. The disadvantage is that this structure is even more memory-consumptive than the non-hierarchical structure. The typical data structure is the complete (balanced) tree.

3. **Hierarchical modeling structure:** In this structure, a tree is built upon the volume elements, and a larger volume element on the next higher level replaces a group of volume elements completely if they are homogeneous.

   Advantages are efficient storage and efficient operations. The main disadvantage is the difficult visualization. The typical *data structure* is the tree.

6.7.2.2 The Local Structure

According to the regularity of the computer internal model one can distinguish the following levels (assuming a 3D world):

1. **Regular in 3 dimensions:** Total regularity is required, no preferred direction is given.
2. **Regular in the view plane dimensions (2 dimensions):** Since particularly the screen operations – like visualization or shading – are generally rather time-consuming, it might make sense to prefer these two dimensions rather than the third[10].
3. **Regular in the depth information (1 dimension):** Here also the two view plane dimensions are preferred to the third (being *non*-regular and only the third is kept regular.
4. **Non regular.**

In general, non-regular spatial enumeration schemes are called *cell decomposition* (see [Carlbom *et al.*, 1985]). In cell decomposition schemes, cells may have an arbitrary number of faces. Using curved cells gives a good approximation of the object to be represented (see Fig. 6.21).

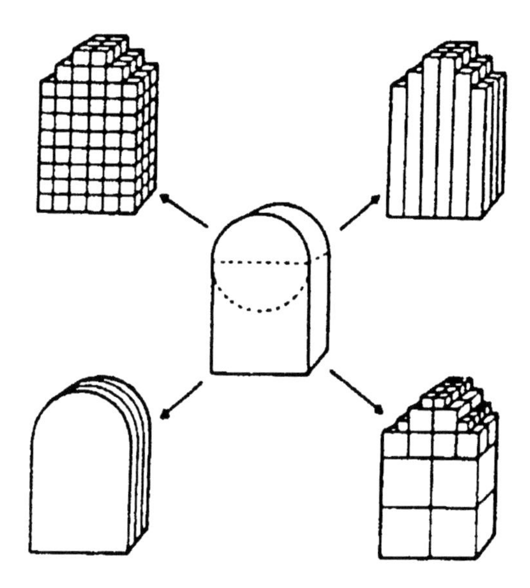

Figure 6.21: Different kinds of cell decomposition schemes

Cell decompositions are unambiguous but in general not unique. Validity may be difficult to establish. Furthermore, cell decompositions are usually neither concise nor easy to create. But one should remark that cell decomposition techniques are valuable for computing certain properties of objects such as mass or center of gravity.

[10]To us it does not seem to make sense to use a model that is regular in one view plane dimension and not regular in the other. If one wants to model in a screen-oriented way, both view plane dimensions are equally important.

6.7.2.3 Classification Scheme

According to the two lists of structuring criteria, one can now give a classification of the main spatial enumeration techniques. Table 6.1 shows such a classification.

Global Structure → **↓ Local Structure**	*Non Hierarchical*	*Redundant Hierarchical*	*Hierarchical*
Regular in All Dimensions (3D)	Voxel	Redundant Octree	Octree
Regular in View plane (2D)	Dexel	HiDex (Hierarchical Dexel)	QuaDex (Extruded Quadtree)
Regular in Depth Information (1D)	Layer	HiLayer (Hierarchical Layer)	–[11]
Non Regular	3D Mesh (Arbitrary Boxes)	Isothetic BSP-Tree	Extended Octrees

Table 6.1: A classification scheme of approximation models according to their global / local structure

6.7.3 Different Spatial Enumeration Techniques

In the following we present a brief characterization of the various spatial enumeration techniques that have been classified using our scheme. We should like to state that whereas some of the modeling schemes have already been well discussed in the literature (e. g. octrees, voxels), others are just at the beginning of being considered interesting and, thus, rather incompletely researched.

Voxel

For creating a voxel model (see Fig. 6.22 for an example), the modeling space is subdivided into equally sized cubes.[12] Each cube is uniquely mapped to a 3D matrix. A matrix component holds the value "volume" if the corresponding cube is fully inside the object that is modeled. Otherwise the matrix component holds the value "no volume".

Redundant Octree

The modeling is the same as for voxels. Additionally, every 8 cubic elements are combined to a hypercube. Such a hypercube holds the information whether the subcubes are heterogeneous or homogeneous with respect to the respective volume value.

[11] No plausible representation of this type has been found.

[12] Other decompositions (e. g., tetrahedral ones) are also used. In general, the type of the grid does not affect the main properties of spatial enumeration schemes.

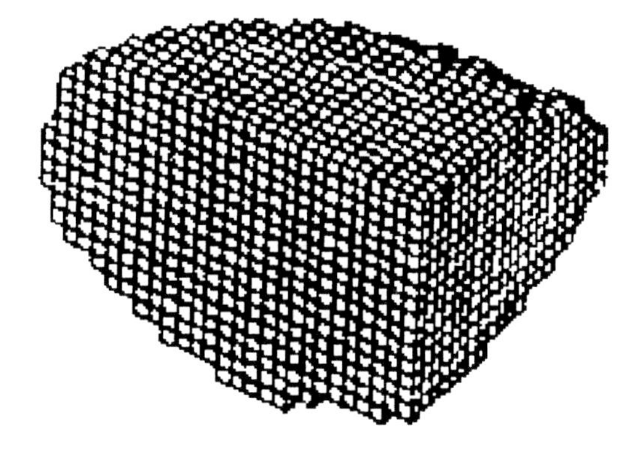

Figure 6.22: Voxel representation

Octree

For octree encoding (see Fig. 6.23 for an example), the modeling space is subdivided into eight identical subcubes. All subcubes are again subdivided if necessary, and so on. A subdivision takes place when a cube is heterogeneous (i. e. the cube lies partially inside and partially outside the object). When a cube is homogeneous or the size of the cube remains smaller than the resolution limit, the cube's volume value is stored in the data structure.

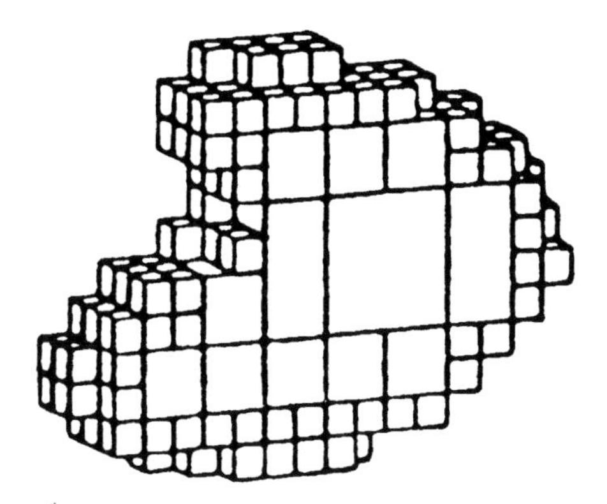

Figure 6.23: Octree representation

For octree encoding, all objects to be modeled are required to have the same dimensionality as the universe. But there are no restrictions on the shape of the objects. Therefore, octrees are the most frequently used spatial enumeration technique.

Using octree encoding, the data is arranged in a hierarchical tree whose nodes represent disjoint cubes of exponentially decreasing size (see Fig. 6.24). Each node has the occupancy status of this region as its property. If the cube is completely contained or not at all contained in the object, then the node is a leaf. In this case it is flagged full (or empty, respectively). Otherwise, only a portion of the cube is occupied by the object. In this case, the node has 8 children, i. e. it points to 8 nodes representing regions that partition the actual cube into

8 subcubes of equal size. By traversing the tree in a proper order, some uniform visiting direction in space can be imposed.

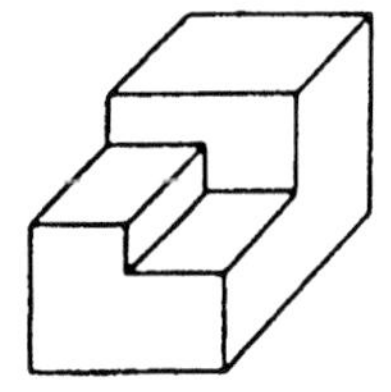

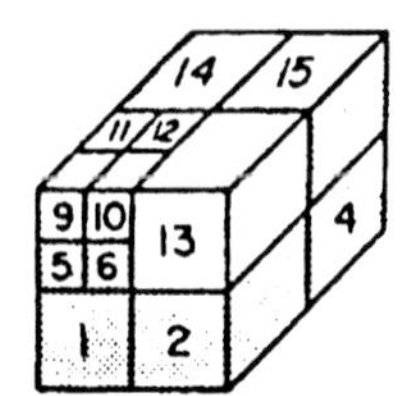

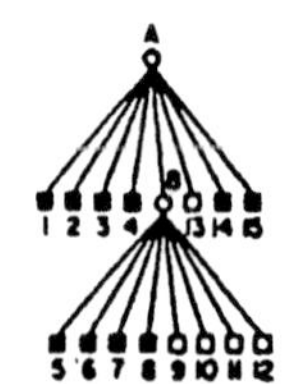

Figure 6.24: Octree encoding

Due to the hierarchical structure, the root represents the entire object. Similarly, the nodes at some level together with the higher nodes completely describe the entire object up to the resolution of that level. Hence, the user has the possibility to trade off computation time against processing precision. This means that it is possible to operate at that level which seems appropriate for the task to be performed. For instance, using a coarse approximation may be very efficient when performing interference checks.

The main drawback of octree encoding is the large amount of data needed for storing curved objects. In [Meagher, 1980], it has been proven that the memory and the processing time required for modeling an object is on the order of the surface area of the object. Depending on the resolution, several million bytes of node storage may be necessary for representing realistic scenes.

Nevertheless, as stated in [Meagher, 1980], this modeling scheme is not that bad if compared to other schemes: One should observe that slight modifications of a modeled object do not cause dramatic changes of the storage needed. For instance, slightly deforming the surface of a sphere will not result in a significant increase of the number of nodes of its octree (as opposed to a CSG system, where it might be necessary to incorporate lots of additional primitives). Furthermore, the high-level parts of the octrees of the original and of the modified object usually will be identical, indicating that the represented objects are similar.

The processing of each node generates 0 to 8 independent sub-calculations. Thus, when using octree-encoding, many computations can be executed in parallel. Algorithmic operations on octrees like, e. g.,

- Boolean operations,
- translations, rotations, reflections, and scaling,
- hidden surface removal,

can be performed by algorithms that can be implemented without using floating point arithmetic or integer multiplication / division.

Dexel

For *dexel modeling* (see Fig. 6.25 for an example), a set of parallel rays is intersected with the object to be encoded. For each ray, the points of intersection with the object are stored in the data structure. For literature, we refer, e. g., to [van Hook, 1986], [Heinzelreiter, 1990], [Mayr and Heinzelreiter, 1991a].

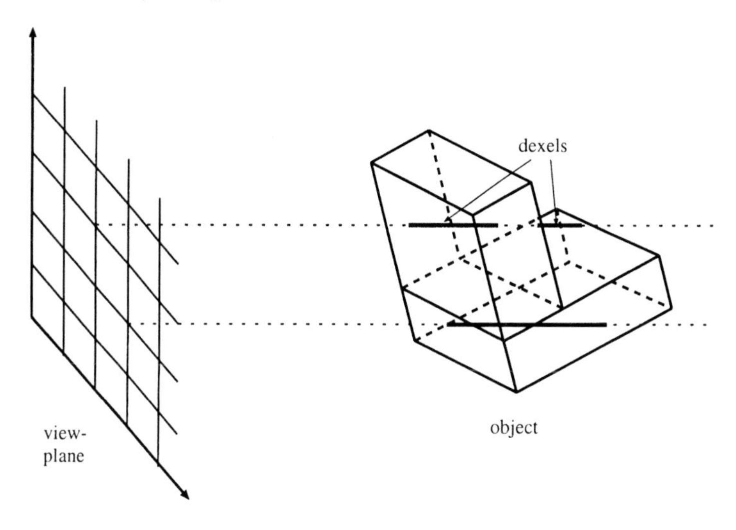

Figure 6.25: Dexel encoding of a (simple) 3D object

Hierarchical Dexel

For *hierarchical dexels* (*HiDex*), the basic structure is the same as for dexels, but each group of four adjunct dexels are (hierarchically) combined into a hyper-dexel.

Extruded Quadtree

For *extruded quadtrees* (*QuaDex*), the modeling space is decomposed into four equally sized regions by two perpendicular planes. If such a region is merely bound by planes we have reached a leave node (all relevant geometric data is stored in the leaf). Otherwise, further decomposition takes place.

Isothetic Layer

The *isothetic layer* model (see Fig. 6.26 for an example) consists of a number of slices that are parallel to the view plane and give a 2D cross section of the object in regular distances to the viewpoint.

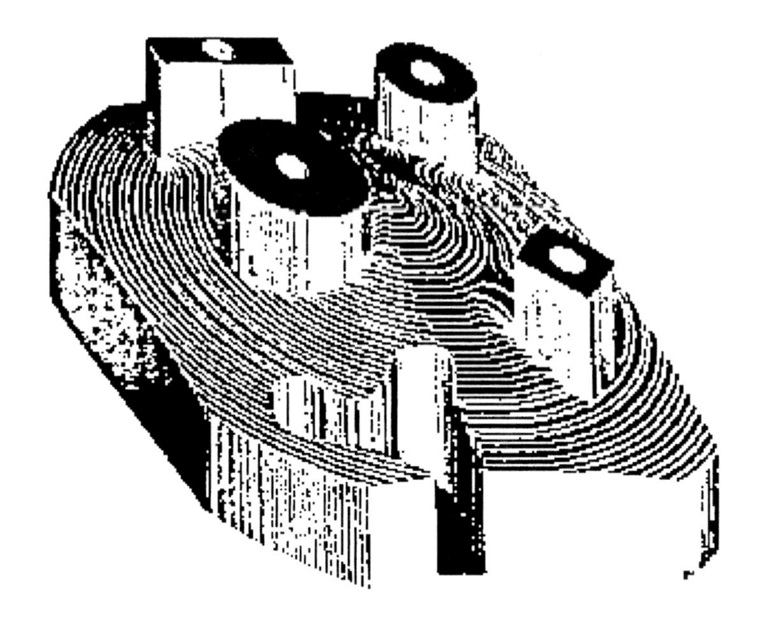

Figure 6.26: Isothetic layer representation

Hierarchical Layer

The *hierarchical layer* modeling (*HiLayer*) is the same as for the layer, but there are layers on the next higher level combining two by two layers (like a conversion into layers using a lower resolution).

3D Mesh

According to the shape of the object, for constructing a *3D mesh* the object is decomposed into a number of boxes (sometimes: isothetic boxes). The vertices of these boxes together with the volume information can be stored in a 3D matrix.

Isothetic BSP-Tree

According to some characteristic edges or vertices of the object, for creating an *isothetic BSP-tree* (binary space-partitioning tree) the modeling space is subsequently binarily subdivided in two, thus getting a tree with volume and non-volume nodes.

The classification as redundant hierarchical / non regular is not totally correct since the hierarchical tree is, basically, not redundant. However, practical applications show that the tree tends to get a very balanced character (reason: infinite separation lines). A BSP-tree is also not completely non-regular, but it is object-dependent and therefore oriented on the coordinate values of the object. Thus, we classify it this way. For literature, we refer, e. g., to [Preparata and Shamos, 1993], [Chin and Feiner, 1989].

Extended Octrees

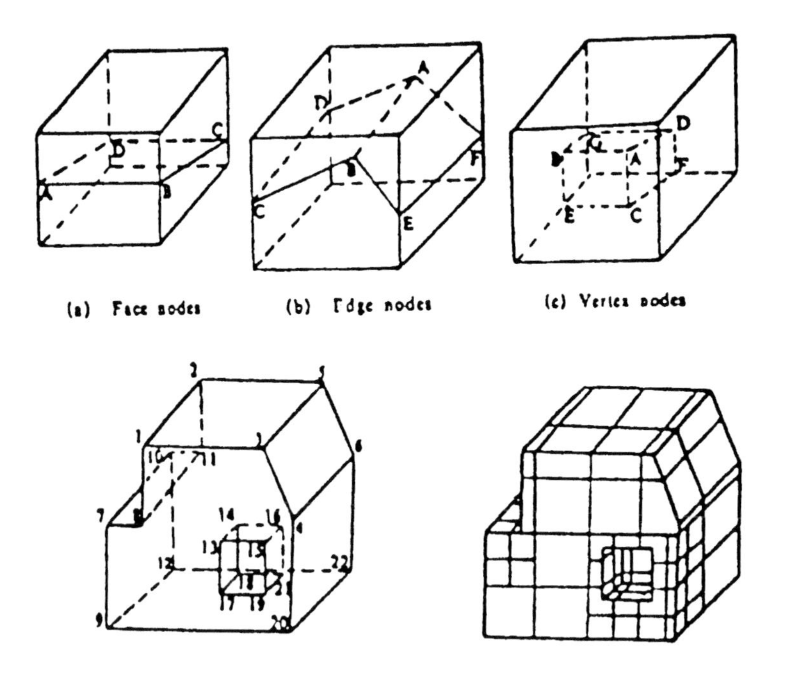

Figure 6.27: Extended octree representation

The standard octree is transferred into an *extended octree* adding three node types[13]:

1. face node,
2. edge node, and
3. vertex node.

This leads to a more accurate and more compact representation, see Fig. 6.27.

The classification as hierarchical / non regular is not totally correct since, basically, the octree is a regular spatial enumeration technique. However, the additional node types are oriented on the faces, edges, or vertices of the object and therefore depend on their coordinate values. Thus, we do the classification this way.

For literature, we refer, e. g., to [Samet and Webber, 1988], [Sun, 1989]. A similar approach to the (exact) handling of polyhedra by means of octree encoding has been presented in [Brunet and Navazo, 1985] and [Navazo *et al.*, 1986].

[13] There is also a different way of extending octrees using non-regular subdivision (see [Samet, 1984] for a good survey on the two-dimensional case). This extension is not discussed here.

6.8 Hybrid and Dual Modeling Schemes

Hybrid (or non-homogeneous) representation schemes are designed by combining two or more of the representations discussed in the previous sections. In this way, the domain of the modeler is enlarged. The following items provide some examples of hybrid schemes.

- **CSG / B-Rep hybrid**: Objects are represented by means of a CSG tree whose leaves are either CSG primitives or boundary representations of non-primitives. This hybrid scheme is of particular importance for the input of geometric objects and for modeling free-form volumes within a CSG scheme.
- **CSG / solid-sweep hybrid**: Objects are represented by means of a CSG tree whose leaves are either CSG primitives or solid-sweep representations of non-primitives. This hybrid scheme is of particular importance for NC verification.
- **Sweep / gluing**: Using the gluing operator is a common method for enriching the domain of a sweep representation.

For the sake of fast algorithms for different application purposes, redundancy is sometimes introduced by storing two (or more) dual representations of one object. Typical examples are provided by the following items:

- **CSG + spatial enumeration, CSG + cell decomposition**: Incorporating these shape approximation techniques into a CSG modeler permits fast calculations of mass properties, for instance.
- **B-Rep + spatial enumeration, B-Rep + cell decomposition**: Again, the space subdivision is often used for calculating mass properties.
- **CSG + polyhedral modeling**: The polyhedral model is often used for displaying the object.
- **CSG + octree, B-Rep + octree**: The octree representation may serve as a coarse approximation of objects. For instance, it enables fast collision checks.
- **CSG + B-Rep**: The ultimate modeling scheme?

6.9 Example: Modeling NC Machines Using the Layer Tree Model

6.9.1 Components of Typical NC Machines

In the following, we concentrate on the modeling of NC machines, because this class of mechanisms already found a broad acceptance in industrial plants, whereas pure robots still occur rather rarely on the shop floor, particularly in small and medium-sized companies.

When analyzing typical NC machines, particularly such for drilling, milling, and turning purposes, one can observe that such a machine is composed of the following components (see Fig. 3.3):

1. body,
2. transmission,
3. carriage (optional),
4. fixture for the tool,
5. tool,
6. fixture for the workpiece,
7. workpiece.

From a geometric point of view, an NC machine consists of the following main components:

- stationary parts (they can be described using geometry only) and
- mobile parts (the links of a kinematic chain; their kinematic behavior must be specified additionally).

In the following, we use the term *NC machine* in such a way that it includes all these components of an NC machine.

A name must be specified for each part of an NC machine that is unique within the specific work cell description. This name can be used for identifying and referencing the single objects and parts of an object. These operations are necessary, e. g., for kinematics assignment, kinematics restrictions, localization of a detected collision, and the identification of an object using the mouse.

For being modeled, all these components are described by their geometry using the *layer tree concept* that will be introduced in the following section (see [Mayr, 1991c] for details).

The specification is designed in such a way that it is possible to generate a description of a geometric model of an NC machine using a suitable geometric editor (e. g., parameterizing an NC machine, see [Lehmann, 1989]).

6.9.2 The Layer Concept

Power of the Model

Our main motivation and goal for developing a new – extended – CSG scheme was the uniform and easy modeling of the following objects:

- extruded 2D contours ("$2\frac{1}{2}$D objects"),
- 2D contours rotated with respect to a 3D line (3D objects symmetrical with respect to a line),
- arbitrary 3D CSG objects (restricted only by the class of primitives allowed).

Our CSG scheme gets this power by utilizing a very general description for the primitives allowed (the *layer* concept, see Chapter 6.9.2) as well as the extended CSG tree (*layer tree*) specified in Chapters 6.9.3 and 6.9.4.

The class of objects that can be described is the same as that of an ordinary CSG scheme, extended by tori and oblique cones. However, the internal tree is much more compact because of the large variety of primitives that is possible. This fact holds particularly for technical objects.

Data Structure of a Layer

Roughly, a *layer* is a solid object delimited by two parallel planes (bottom and top plane) and the lateral area that joins the two planes. Additionally, objects that are symmetrical with respect to a line can also be described as a layer. Fig. 6.28 shows some examples of objects that can be represented as layers.

We see from the examples in Fig. 6.28 that we need two contours (in the top and bottom plane, respectively) and a height for a complete description of the geometry of a layer. The border of the layer is composed of n surface patches where n is the number of elements of the contours. (If the two contours have different length, the shorter one is thought to be completed by repeatedly adding the end point of that contour until the length of the longer contour is reached.) Each surface patch is defined by an element of the bottom contour and its corresponding element of the top contour.

The two contours are stored in the arrays `bot-contour` and `top-contour`. They always have to be simple (no intersections of two contour elements) and closed (start point and end point are equal). The bottom contour is specified in the XY-plane of the coordinate system of the layer. The `height` can be any positive or negative real value.

The position and the orientation of a layer with respect to the parent object[14] as well as additional information about the object (color, technological data, ...) is stored in the parent object. The contour description is given as a list of `Triplets`, where the first element is an information code and the last two represent the coordinates of a point (see [Mayr, 1987]). Thus, the data structure for a layer can be specified in the following way:

```
struct  Layer
{
        Triplet     bot-contour[]
        Triplet     top-contour[]
        double      height            – positive or negative
}
```

Up to now, the contour elements are comprised only of line segments and circular arcs. If the demand arises, it is possible to add other contour primitives, e. g., the primitive element spline, to this contour description.

[14] The *parent object* is the father of the object node in the layer tree, see Chapter 6.9.3.

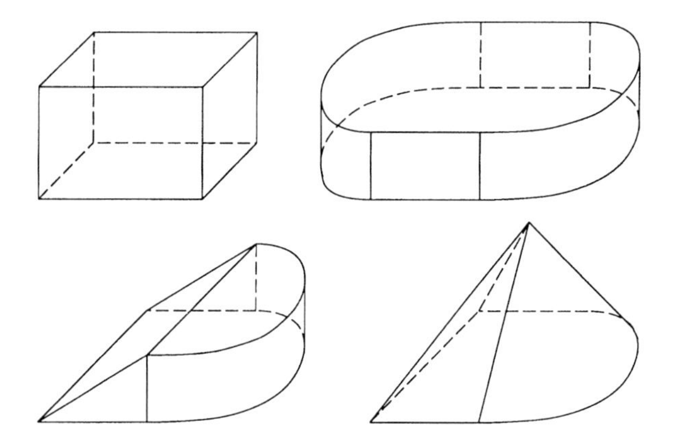

Figure 6.28: Objects represented as layers

Semantics of a Valid Layer

Layers have to be specified in such a way that they are composed only of uniquely defined primitive elements (in order to prevent free form surfaces). In order to guarantee this condition, the lateral area of a valid layer (the area that connects the two contours) may be only comprised of the following elements:

- plane quadrilateral areas,
- parts of the lateral area of oblique cylinders,
- parts of the lateral area of oblique cones.

See Fig. 6.29 for examples.

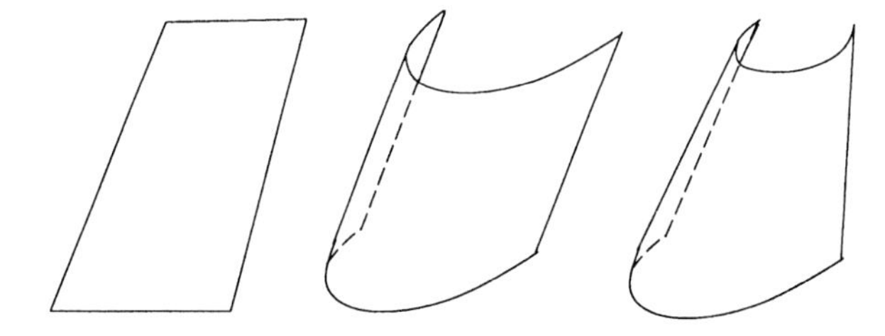

Figure 6.29: Primitive elements of the lateral area of a valid layer

If the boundary of an object is composed of only two parallel contour areas and primitive elements (as defined above) then it is called a *valid layer*.

Such valid layers are of interest, because existing geometric 2D data bases in industry often use contours as a primitive element. Therefore, 3D information has to be generated using these data bases converted into our layer model. This is very important from an economic point of view because of the availability of huge geometric 2D data files that already exist. In this way they can be re-used for 3D purposes, too.

Moreover, objects that are symmetrical with respect to an axis are also considered as valid layers. In this way, spheres and tori, for instance, are also primitives available within our modeling scheme.

Conditions for a Valid Layer

Obviously not all combinations of top and bottom contours constitute a valid layer (see Fig. 6.30). In order to assure that two contours with an equal number of elements define a valid layer, two corresponding elements of the contours (i. e. the ith element of the bottom contour and the ith element of the top contour) have to fulfill the conditions enumerated below:

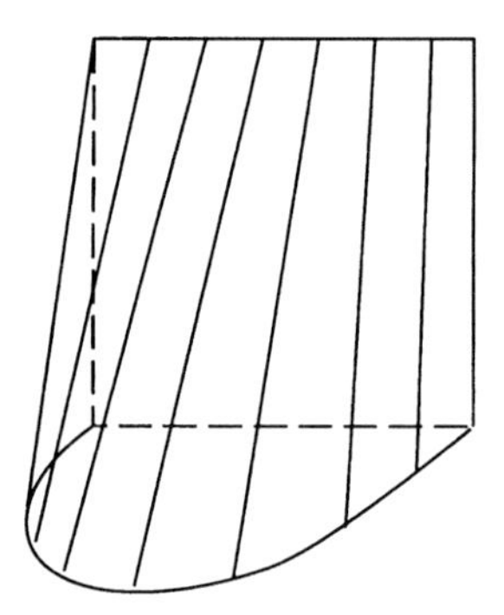

Figure 6.30: An invalid layer

1. Points are mapped to points, straight lines or arcs.
2. Straight lines are mapped to points or straight lines. If a straight line is mapped to a straight line, the four points that define the two straights have to be coplanar.
3. Arcs are mapped to points or arcs. If an arc is mapped to an arc, the two arcs must have the same orientation and the start points and the end points have to be coplanar. Also, the angles of the arcs must be identical in order to get a boundary of that part of the lateral area that consists only of straight lines and circular arcs and can, thus, be adjoined correctly with the adjacent parts of the lateral area.

Efficient Representation of Symmetrical Objects

The most commonly used objects in industry are symmetrical objects (in turning processes) and objects with identical or partially identical contours. In order to be able to represent these objects efficiently, the layer model is extended in the way described in the following two sections.

Objects that are symmetrical with respect to a line are uniquely defined by a contour and an axis of revolution (see Fig. 6.31). Such an object can also be stored in the layer data structure. The defining contour is stored as the bottom contour and the axis of revolution is stored as the top contour consisting of two points. A height of zero indicates that the layer is to be interpreted as a symmetrical object.

The layer data structure of a symmetrical object has the following form:

```
Layer    RotSymmObject
{
         bot-contour[]       - the defining contour
         top-contour[]       - the axis of revolution
         0                   - characterizes a symmetrical object
}
```

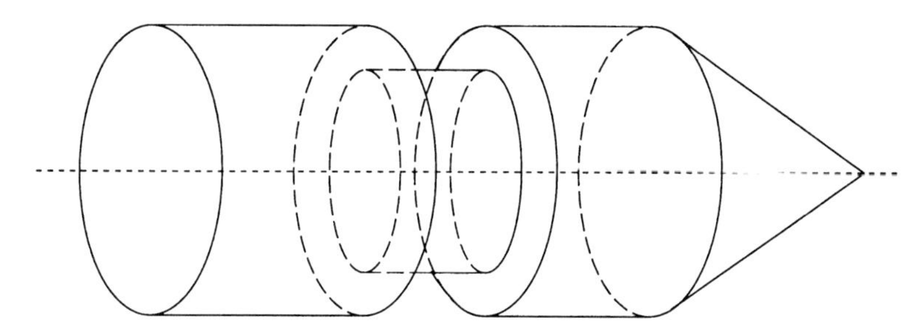

Figure 6.31: An object symmetrical with respect to an axis

Efficient Representation of Prismatic and Conic Objects

In technical applications, frequently general prisms (arbitrary $2\frac{1}{2}$D objects) and general cones have to be handled (see Fig. 6.32). General prisms are characterized by a top contour that is identical to the bottom contour. General cones are characterized by one contour containing less elements than the other. When using our standard layer representation we have to store a lot of redundant information for these special objects. In order to avoid this inefficiency we extend our model in the following manner:

- ***General Prism:***

  Since the two contours are identical, only the item `bot-contour` is stored. The length of the top contour is equal to zero.

  The layer data structure of a general prism has the following form:

```
Layer    GeneralPrism
{
         bot-contour[]       - (= top-contour)
         []                  - top-contour contains only an end marker
         height
}
```

- ***General Cone:***

  When the lengths of the top contour and the bottom contour are different, the last point (end point or single point) of the shorter contour is mapped to the remaining elements of the longer contour.

  The layer data structure of a general cone has the following form:

```
Layer    GeneralCone
{
         bot-contour[]        - bot-contour and
         top-contour[]        - top-contour have different lengths
         height
}
```

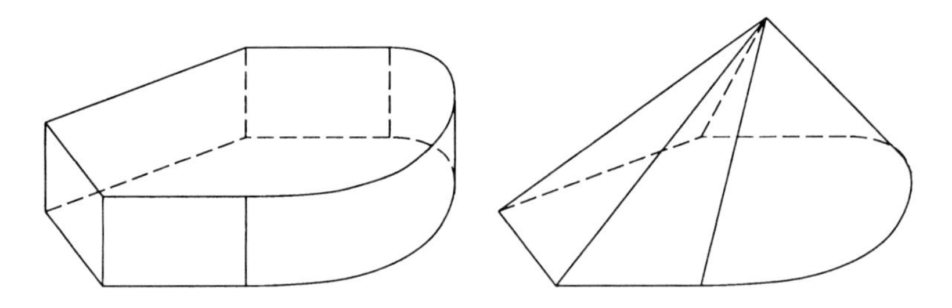

Figure 6.32: General prisms and cones

6.9.3 The Layer Tree

The layer tree is an efficient extension of the CSG tree for representing solid models composed of additive and subtractive volume primitives.

A *layer tree* describing a geometric object consists of a root node that is of type `SolidObject` and a number of internal nodes and leaf nodes that are of type `Element`. For details on the structure of these nodes see Chapter 6.9.4. Each element node of the tree can have an additive volume value (additive component, *union*) or a subtractive volume value (subtractive component, *difference*). The third CSG operation, the *intersection*, can be gained using union and difference.

The following facts make the layer tree superior to the standard CSG tree:

- Although there is only one basic primitive (the *layer*), this primitive allows the modeling of, both, objects that are rather natural for the NC programmer (e. g., objects gained by extruding a closed 2D contour into 3D) and objects constructed of rather general primitives like truncated oblique cones or non-convex oblique prisms.

- Due to the powerful basic primitive, a layer tree is generally more compact than a CSG tree for the same geometric object. This results in less memory consumption and faster computation times for transforming objects into a different representation.

- The layer tree is stored internally as a binary tree. This allows a very efficient processing of the geometric objects stored. However, for the designer one object (or sub-object, called element by us) is not restricted to be constructed of only two elements or primitives. So the programmer can, for instance, specify a block with six holes exactly in that way. He is not forced to create artificial objects by combining two holes to a new element only to fulfill the pure binary tree structure. An example how to specify a layer tree for a given object (Fig. 6.33) is presented in Figs. 6.34 and 6.35.

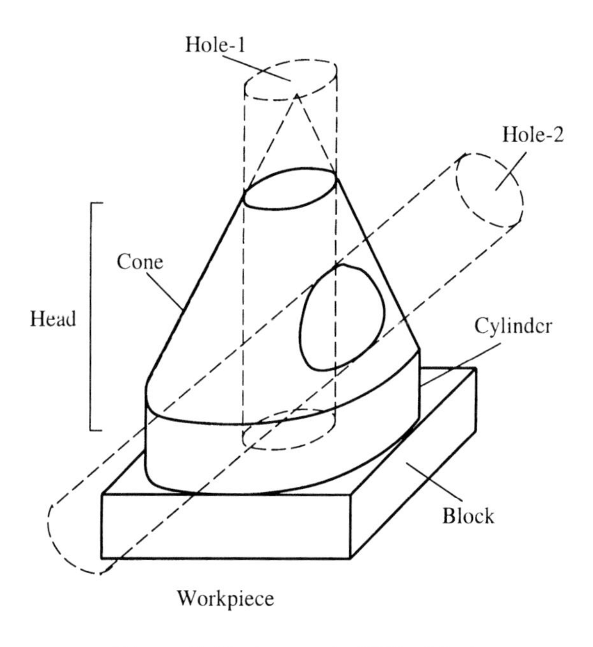

Figure 6.33: A sample CSG object

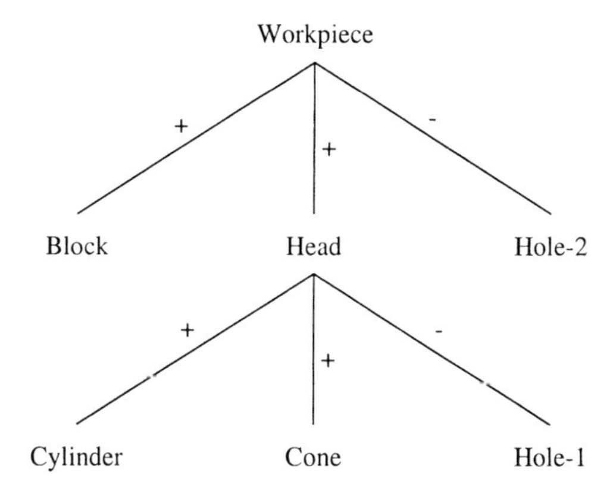

Figure 6.34: The input layer tree for the sample object

Since the standard CSG tree is only a special case of the layer tree, our representation is compatible with any standard solid representation using CSG trees.

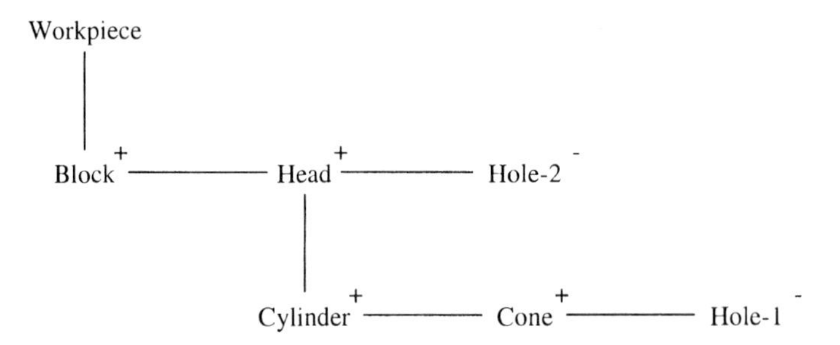

Figure 6.35: The internal layer tree for the sample object

One should note that the layer tree is – like the standard CSG tree – not a unique representation of a geometric object. For correct object retrieval from the layer tree, one has to process the tree using a pre-order traversal algorithm, such that the tree is traversed depth-first and from left to right.

6.9.4 Nodes of the Layer Tree

The Root Node

The root node describing the `SolidObject` consists of the following entries:

```
struct  SolidObject
{
        string    name        - object descriptor
        Element   *object     - geometric information
        Frame     position    - position of the object (world coordinates)
        string    technData   - technological information
}
```

Notes:

1. The object descriptor must be unique within one data base such that it can be used for referencing.
2. The geometry pointer `object` specifies the first element node.
3. The frame describing the position is specified by a 4×4 matrix[15].
4. The technological information might for instance contain information about the material (specifying the color).

[15] In affine geometry, a 3×4 matrix can be used for implementation since the fourth vector always is (0 0 0 1).

Internal Nodes and Leaves

The root node describing an `Element` consists of the following entries:

```
struct  Element
{
        string      name         - element descriptor
        {add,sub}   volValue     - additive or subtractive volume value
        Layer       geometry     - geometry primitive
        Polytope    enclosure    - enclosing polytope for collision checking
        Frame       position     - position relative to the father node
        string      technData    - technological information
        Element     *brother     - element on the same tree level
        Element     *son         - element on the next lower tree level
}
```

Notes:

1. The element descriptor must be unique within the corresponding `SolidObject` such that it can be used for referencing (e. g., the position frame). The element descriptor may be empty if the element is not to be referenced.
2. The geometry of the element is specified by a layer primitive. The structure of a layer is specified in detail in Chapter 6.9.2.
3. A procedure to enclose a layer by a polytope as well as collision checking strategies for such polyhedral objects are presented in [Mayr, 1991b].
4. The frame describing the position is specified by a 4×4 matrix. The father node can be an `Element` node or a `SolidObject` node.

Artificial Element Nodes

Artificial elements are used for components of the solid object that belong together logically, but only consist of the volume of the elements they are composed of. This means that the node characterizing the artificial element does not contain any layer information describing the geometric volume of the node. Instead, only the layer information of the sons of this node describes the volume of this node. To get a valid volume information, the first son that has a non-empty layer must have an additive volume value.

Characteristics of artificial element nodes are that `layer` has to be empty, whereas `*son` must not be empty.

One can easily see that any artificial element node (i. e. the node of an artificial element) can be eliminated by rearranging the layer tree. However, the structure of the solid model is then, in general, no longer explicitly retrievable from the tree. Furthermore, the manual rearrangement of the tree might easily lead to errors and does not give much storage advantage.

An artificial element node is contained in the layer tree of Fig. 6.34.

6.9.5 Modeling the Geometry of NC Machines

An NC machine geometry is composed of stationary parts together with the links and joints of the mechanisms. The parts and the links can be represented using specific object primitives. Using the concept of CSG, the machine is described by the union of its primitives. The CSG scheme is restricted by the fact that only the *union* operation is allowed for the modeling process (see Chapter 6.9.6 for the reasons). However, the single primitives need not necessarily be disjoint.

The models used for describing an NC machine are determined by the input interface (*input model*), efficiency of the simulation operations (*internal model*), and fast visualization (*output model*). All three models are hierarchically structured.

In conformity with the model for workpieces, tools, and fixtures, a slight extension of the *layer tree* concept is used for describing the NC machine. By doing so, a unique representation of all necessary components for machining simulation and manufacturing simulation is made possible. Additional information for the single parts of the machine is stored in the *technological information* fields of the nodes of the layer tree.

The layer tree concept is customized for describing the NC machine in the following way:

1. A *layer tree* consists of a root node (*object*) and a number of internal nodes and leaf nodes (*element*). For an NC machine description, each element node of the tree can only have an additive volume value (*union* operation). See Chapter 6.9.6 for the reasons of this restriction.
2. The layer tree is also used for representing the hierarchy of the internal model of the NC machine. To enable this, a pointer to the internal (polyhedral) model of each element of the machine description is stored in the respective node of the layer tree. Consult the next section for details on this extension.

The manufacturing simulation concentrates on the task within a work cell, i. e. the motions of the mechanisms in the object world. The impact on the workpiece (machining process) is not simulated. Therefore, the main operation on the internal model of the work cell is the visualization of the single mechanisms and objects at their actual position at specific time moments (snapshots at those moments). This influences the choice of the suitable internal model of the object world considerably.

All 3D graphic standards that are currently available do only support the visualization of polyhedral objects. Consequently, we represent our objects by a polyhedral enclosure of their boundary. (As Chapter 6.9.6 shows, the use of polyhedral enclosures instead of just polyhedral approximations is inevitable for a correct collision detection.) For each single object we store the polygonal boundary of each facet of the enclosure. However, we also keep the facet information, i. e. the information about the interior and the exterior, in order to enable a visualization using hidden line techniques or shading techniques. The polyhedral information for each element of the NC machine description is added to the respective object node in the layer tree. This concept also enables the addition of hierarchically structured convex hulls for the objects in order to allow different levels of detail and speed up collision checking.

For approximating an object whose contour description contains arcs, an exactness factor has to be specified. Naturally, the approximation is the better the smaller the exactness factor is. However, one should keep in mind that the visualization slows down with the increasing number of facets to draw. If the visualization is done using wire frames, a huge number of lines may confuse the user instead of giving him additional information.

The layer tree concept allows the additional storage of technological information to each of the elements of which an object is composed (i. e. to each of the nodes of the tree. This field can be used to indicate the technological purpose of the component (e. g. fixing device, sliding carriage, or axis). In addition to the name of the component this information can be used for the interactive analysis.

6.9.6 Modeling Conditions and Restrictions

The following conditions / restrictions have to be obeyed when modeling an NC machine in a work cell:

- The only operation that is allowed for composing an object using primitives is the *union* operation. The main reasons for this restriction are:

  1. better chances for an accurate collision detection,
  2. easy and efficient conversion into a wire frame model, and
  3. better suitability for the identification and interactive analysis using the mouse.

  However, holes can be specified since our layer model allows the definition of non-convex objects (see the next item).

- Non-convex objects can be easily defined using our layer scheme. However, as the automatic decomposition of a non-convex solid into convex ones takes time and may cause a strange appearance of the object in wire frame representation, we recommend the use of convex elements whenever this is possible.

- Objects that contain arcs in their layer description are enclosed by polyhedral facets. It is not sufficient to just approximate them by choosing vertices for the facets that lie on the boundary of the object. Only if the polytope is a real enclosure of the object, this will guarantee conservatism when using this polytope for collision checking. (Conservative checking means that in situations, where two objects are close to each other, the collision checking algorithm reports collision rather than no collision. Consequently, if no collision is reported, it can be guaranteed that also the enclosed objects do not collide.) Although the polyhedral enclosure is done automatically, the designer should keep the concept of enclosing the objects in mind. For this reason, two moving objects containing arcs in their boundary should pass each other at some security distance.

- Arcs that are contained in an object description should not be resolved in advance. If some arcs are already approximated by lines in the input and others are not, this may lead to a huge number of facets and result in a chaotic wire frame visualization.

Chapter 7

Kinematic Models

The goals of this chapter are getting familiar with forward (direct) kinematics and inverse (backward) kinematics of a mechanism, being able to determine its Denavit-Hartenberg parameters, and understanding the process for determining the kinematics model of the mechanism.

7.1 Definition of a Mechanism

Any mechanism (robot, manipulator, NC machine, see Fig. 7.1) can be considered to consist of a series of *links* and *joints*, similar to a human arm and hand, see Fig. 7.2. Links are the mobile components of a mechanism, two or more links are each connected by means of joints. Depending on whether such a joint allows translational motion or rotational motions, it is called *rotational joint* (*revolute joint*) or *translational joint* (*prismatic joint*), respectively. At the end of such a kinematic chain, normally a gripper (in case of a robot) or a tool (in case of an NC machine) is placed. Therefore, the last frame of a kinematic chain is generally called *gripper frame* (*end effector*) or *tool mounting frame*.

In the following, we will treat mechanisms that contain only *open kinematic chains*, i. e., each joint has to connect exactly two links and there must not be any loop in the kinematic chain. Hence, kinematic schemes such as parallelepipeds are not further treated in this section. The following chapter loosely follows the introduction given in Paul's landmark book [Paul, 1981].

7.2 The Forward Kinematics of a Mechanism

7.2.1 Describing Kinematic Chains Using Sequences of Coordinate Frames

Suppose that a coordinate frame is embedded in each link of the mechanism. Then it is possible to specify the position and orientation of link $n+1$ with respect to link n by means of a homogeneous transformation. Historically, the homogeneous transformation describing the relation between one link and the next has been called an *A matrix*, see [Denavit and Hartenberg, 1954]. Hence, for a mechanism consisting of m links the position of the end

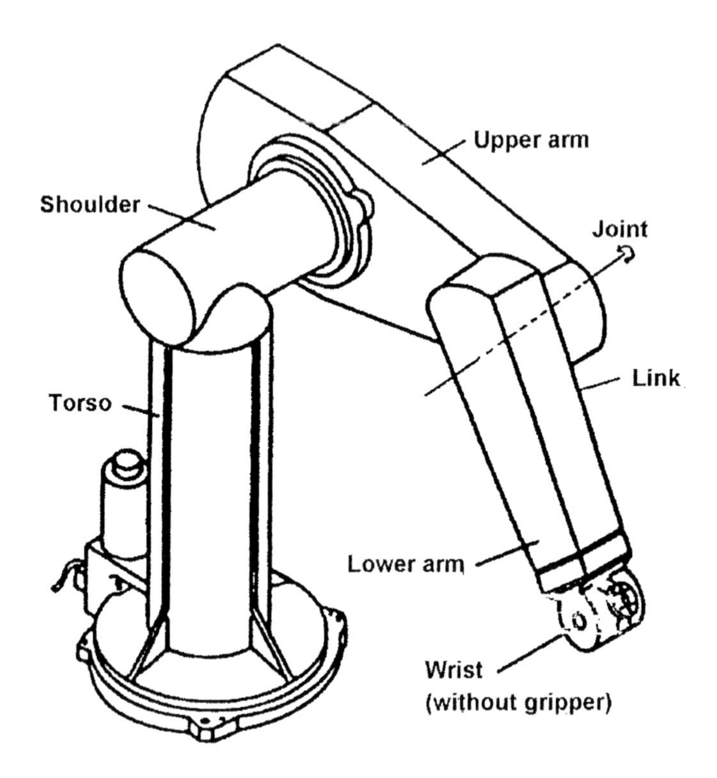

Figure 7.1: Components of a mechanism (robot)

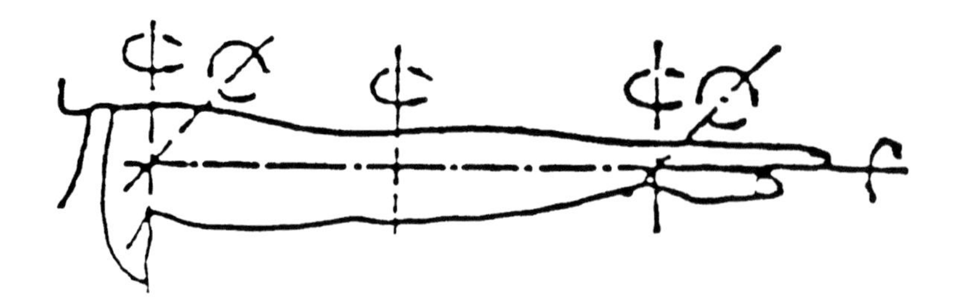

Figure 7.2: The kinematics (motility) of the human arm

effector with respect to the base frame is given by

$$T_m = A_1 \cdot A_2 \cdot \dots \cdot A_m,$$

where A_i defines the position and the orientation of link i relatively to link $i-1$.

Historically, this product of A matrices has been called a T *matrix*. By means of approach vector, orientation vector, and normal vector (forming a right-handed set of vectors, see Fig. 7.3 [Paul, 1981]), the T matrix can be written as

$$T_m = \begin{bmatrix} n_x & o_x & a_x & p_x \\ n_y & o_y & a_y & p_y \\ n_z & o_z & a_z & p_z \\ 0 & 0 & 0 & 1 \end{bmatrix}.$$

Thus, the position and orientation of the gripper can be specified by means of 12 variables. Whereas p is not subject to any restriction, the vectors n, o, a have to fulfill the equations

$$\begin{aligned} \|n\| = \|o\| = \|a\| &= 1, \\ n \cdot o &= 0, \\ n \cdot a &= 0, \\ o \cdot a &= 0, \end{aligned}$$

due to the orthonormality of the coordinate frame. Since 6 equations are imposed on 12 variables, 6 variables remain independent. These 6 variables ("degrees of freedom") are necessary in order to describe an arbitrary frame (i.e., position and orientation) in 3D.

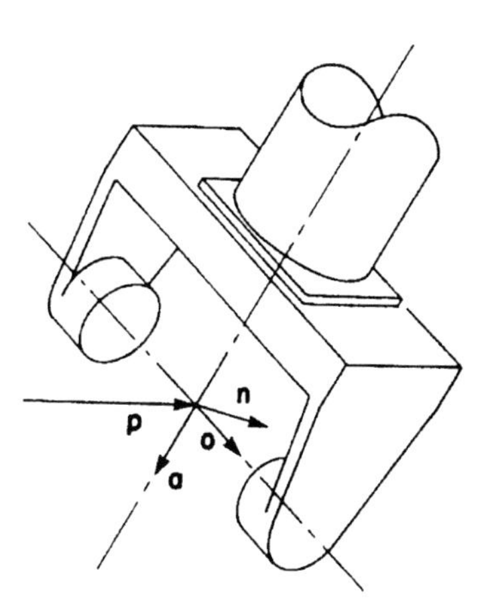

Figure 7.3: n, o, a, and p vectors

7.2.2 Alternatives for Specifying the Orientation of the End Effector

Euler Angles

Euler angles describe any possible orientation in terms of a rotation ϕ about the z axis, then a rotation θ about the new y axis, y', and finally, a rotation about the new z axis, z'', of ψ.

Hence, see Fig. 7.4,

$$\text{Euler } (\phi, \theta, \psi) = \text{Rot } (z, \phi)\text{Rot } (y, \theta)\text{Rot } (z, \psi).$$

Expanded, the Euler rotation is given by

$$\text{Euler } (\phi, \theta, \psi) = \begin{bmatrix} \cos\phi\cos\theta\cos\psi - \sin\phi\sin\psi & -\cos\phi\cos\theta\sin\psi - \sin\phi\cos\psi & \cos\phi\sin\theta & 0 \\ \sin\phi\cos\theta\cos\psi + \cos\phi\sin\psi & -\sin\phi\cos\theta\sin\psi + \cos\phi\cos\psi & \sin\phi\sin\theta & 0 \\ -\sin\theta\cos\psi & \sin\theta\sin\psi & \cos\theta & 0 \\ 0 & 0 & 0 & 1 \end{bmatrix}.$$

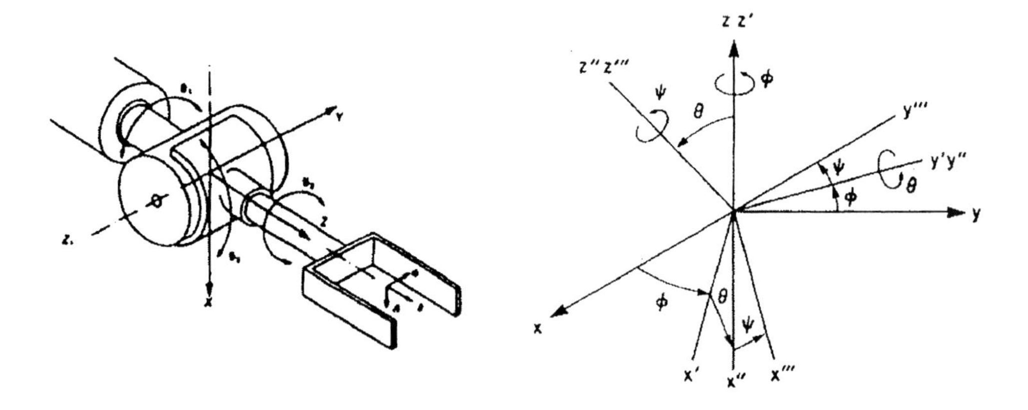

Figure 7.4: Euler angles

Roll, Pitch, and Yaw

Roll, Pitch, and Yaw angles describe any possible orientation in terms of a rotation ψ about the x axis, then a rotation θ about the y axis, and finally, a rotation ϕ about the z axis. Hence, see Fig. 7.5 [Ránky and Ho, 1985],

$$\text{RPY } (\phi, \theta, \psi) = \text{Rot } (z, \phi)\text{Rot } (y, \theta)\text{Rot } (x, \psi).$$

Expanded, the RPY rotation is given by

$$\text{RPY } (\phi, \theta, \psi) = \begin{bmatrix} \cos\phi\cos\theta & \cos\phi\sin\theta\sin\psi - \sin\phi\cos\psi & \cos\phi\sin\theta\cos\psi + \sin\phi\sin\psi & 0 \\ \sin\phi\cos\theta & \sin\phi\sin\theta\sin\psi + \cos\phi\cos\psi & \sin\phi\sin\theta\cos\psi - \cos\phi\sin\psi & 0 \\ -\sin\theta & \cos\theta\sin\psi & \cos\theta\cos\psi & 0 \\ 0 & 0 & 0 & 1 \end{bmatrix}.$$

Rotation with respect to an Axis

Alternatively, the orientation of the end effector can be specified as a rotation θ with respect to an axis k using the generalized rotation matrix, $\text{Rot}(k, \theta)$. Unfortunately, it is not obvious, which axis of rotation should be chosen in order to achieve some desired orientation.

Other ways of describing the orientation of the end effector are, e. g., cylindrical or spherical coordinates.

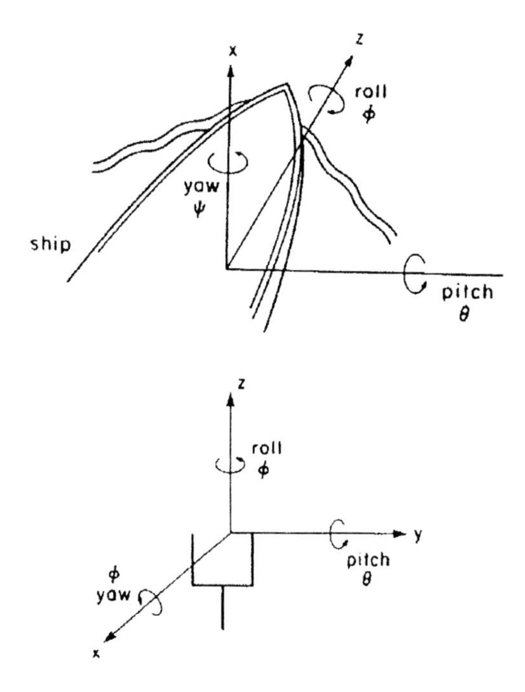

Figure 7.5: Roll, pitch, and yaw angles

7.2.3 Specification of Denavit-Hartenberg Matrices

7.2.3.1 The Denavit-Hartenberg Parameters

Following [Denavit and Hartenberg, 1954], any link can be characterized by means of two dimensions: the common normal distance a_n, and the angle α_n between the axes in a plane normal to a_n (i. e., this plane is parallel to the two joint axes), see Fig. 7.6 [Paul, 1981]. It is customary to call a_n the *link length* and α_n the *link twist n*.

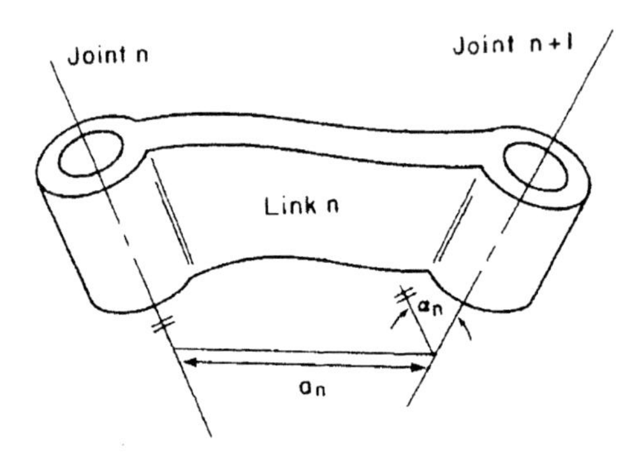

Figure 7.6: Length a_n and twist α_n of link n

Each joint axis has two normals to it, one for each link. The relative position of two links connected in this way is given by the distance between the normals along the axis of joint n, d_n, and by the angle θ_n between the normals measured in a plane normal to the axis, see Fig. 7.7. Usually, d_n and θ_n are called *joint offset* and the *joint angle*, respectively.

These four parameters,

1. the link length a,
2. the link twist α,
3. the joint offset d, and
4. the joint angle θ,

are conventionally called *Denavit-Hartenberg parameters*, due to the landmark paper [Denavit and Hartenberg, 1954].

In order to describe the relationship between the links, coordinate frames are assigned to each link in the following way:

- The origin of the coordinate frame is set to the intersection of the axis of joint $n + 1$ and the common normal (a_n) between the joint axes n and $n + 1$.
- The z-axis of the coordinate frame is aligned with the joint axis $n + 1$. Its x-axis is aligned parallel to a_n. The y-axis is chosen such that these axes form a right-handed set of vectors.

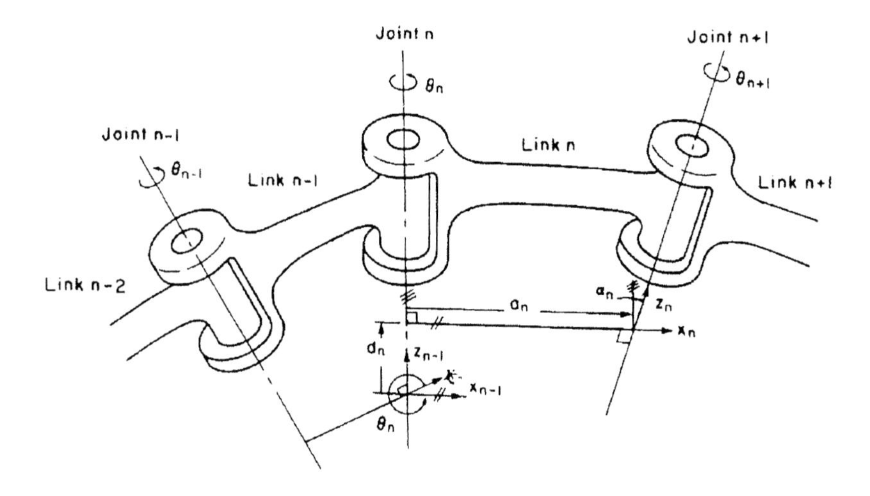

Figure 7.7: Link parameters a, α, d, θ

- The coordinate frame for the origin ("link 0") is defined by the intersection of the axes x_0 and z_0, where x_0 is a vector normal to $z_0 z_B$ and z_0 is defined by the first joint axis.
- In case that the reference coordinate system of the mechanism is different from the (geometric) base coordinate system, the connection between these two coordinate systems has to be described by means of a (fixed) transformation frame (base matrix B).
- The origin of the gripper frame ("link m") is placed coincident with the origin of link $m-1$.
- If the gripper (or tool) mounting frame is different from the position or orientation of link m, the connection between these two coordinate systems has to be described by means of a (fixed) transformation frame, too (gripper / tool center matrix C).

7.2.3.2 Special Conventions for Translational Joints

For correctly modeling translational joints (see Fig. 7.8), the following rules have to be obeyed:

- The joint axis is parallel to the direction of the translational motion of the joint.
- The length of the link (a_n) has no meaning for a translational joint. By definition, a_n is set to 0.
- The position of the coordinate frame is not defined by the translational joint itself; the origin of the coordinate frame is placed at the next defined frame position (base frame, rotational joint, gripper frame).
- The orientation of the coordinate frame is defined by the joint axis (z-coordinate) and by the cross product of z_n and z_{n-1} (x-coordinate).

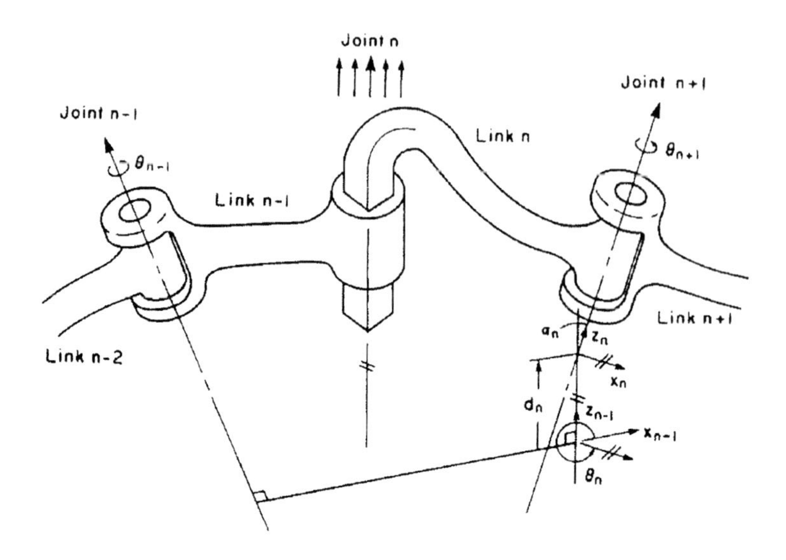

Figure 7.8: Special conventions for translational joints

7.2.3.3 Composing the Denavit-Hartenberg Matrix

The relationship between successive frames $n-1$ and n can be established by the following rotations and translations:

1. Rotate about z_{n-1} by the joint angle θ_n,
2. translate along $z_n = z_{n-1}$ by the joint offset d_n,
3. translate along $x_n =$ rotated x_{n-1} by the link length a_n,
4. rotate about x_n by the link twist α_n.

This relationship can be expressed by the following matrix, the so-called *A matrix* (also-called *Denavit-Hartenberg matrix*):

$$A_n = \text{Rot}\,(z, \theta_n)\ \text{Trans}\,(0, 0, d_n)\ \text{Trans}\,(a_n, 0, 0)\ \text{Rot}\,(x, \alpha_n).$$

Expanded, we learn that

$$A_n = \begin{bmatrix} \cos\theta_n & -\sin\theta_n\cos\alpha_n & \sin\theta_n\sin\alpha_n & a_n\cos\theta_n \\ \sin\theta_n & \cos\theta_n\cos\alpha_n & -\cos\theta_n\sin\alpha_n & a_n\sin\theta_n \\ 0 & \sin\alpha_n & \cos\alpha_n & d_n \\ 0 & 0 & 0 & 1 \end{bmatrix},$$

i. e., the coordinate frame of a link is uniquely defined by the four Denavit-Hartenberg parameters of the link.

By the design of a mechanism, the link length a and the link twist α are already fixed. For a rotational joint, the joint angle θ is a variable and the joint offset d is constant; for a translational joint, d is a variable and θ is constant.

7.2.4 The Problem of Forward Kinematics

Remember that the frame of the end effector with respect to the base is given by the T matrix

$$T_m = B \cdot A_1 \cdot A_2 \cdot \cdots \cdot A_m \cdot C,$$

where B and C are the (optional) constant matrices describing the relation of the kinematic chain to the (geometric) base and the (geometric) gripper / tool center frame. By plugging the D-H parameters into this equation, the position and orientation of the end effector can be evaluated. Thus, we have solved the following

Problem of Forward (or Direct) Kinematics

Given: $a_1, \cdots, a_m$,
$\alpha_1, \cdots, \alpha_m$,
$d_1, \ldots, d_m$,
$\theta_1, \cdots, \theta_m$
(i. e. the D-H parameters describing the joint values and the constant parameters of a mechanism).

Find: T_m
(the position and orientation of the end effector with respect to the base frame).

7.2.5 Example: Mechanism with Two Rotational Joints

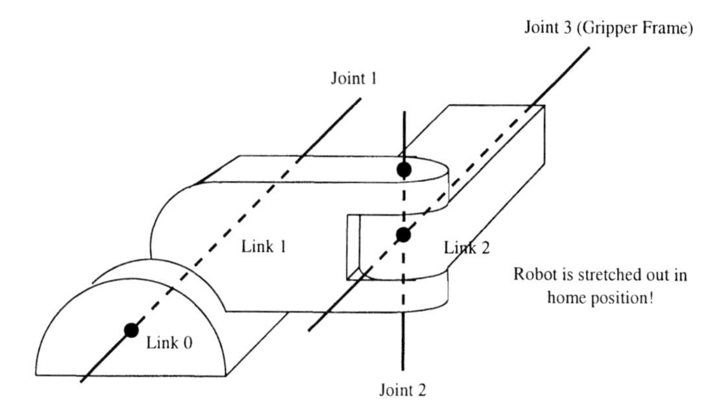

Figure 7.9: Sketch of the mechanism

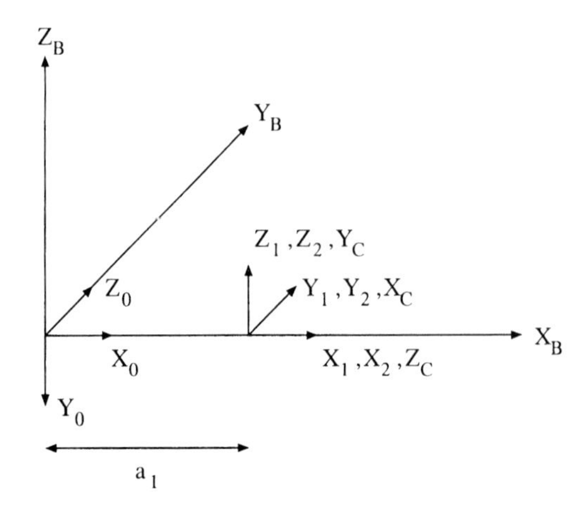

Figure 7.10: Coordinate frames for the joints of the mechanism

Denavit-Hartenberg Parameters

Link	a	α	d	θ
1	a_1	$\frac{\Pi}{2}$	0	θ_1
2	0[1]	0	0	θ_2

Matrices

$$A_1 = \begin{pmatrix} \cos\theta_1 & 0 & \sin\theta_1 & a_1\cos\theta_1 \\ \sin\theta_1 & 0 & -\cos\theta_1 & a_1\sin\theta_1 \\ 0 & 1 & 0 & 0 \\ 0 & 0 & 0 & 1 \end{pmatrix}.$$

$$A_2 = \begin{pmatrix} \cos\theta_2 & -\sin\theta_2 & 0 & 0 \\ \sin\theta_2 & \cos\theta_2 & 0 & 0 \\ 0 & 0 & 1 & 0 \\ 0 & 0 & 0 & 1 \end{pmatrix}.$$

$$B = \begin{pmatrix} 1 & 0 & 0 & 0 \\ 0 & 0 & 1 & 0 \\ 0 & -1 & 0 & 0 \\ 0 & 0 & 0 & 1 \end{pmatrix}.$$

[1]By definition, a_2 goes into the C matrix.

$$C = \begin{pmatrix} 0 & 0 & 1 & a_2 \\ 1 & 0 & 0 & 0 \\ 0 & 1 & 0 & 0 \\ 0 & 0 & 0 & 1 \end{pmatrix}.$$

Since $T_2 = B\ A_1\ A_2\ C$,

$$T_2 = \begin{pmatrix} -\cos\theta_1 \sin\theta_2 & \sin\theta_1 & \cos\theta_1 \cos\theta_2 & a_2 \cos\theta_1 \cos\theta_2 + a_1 \cos\theta_1 \\ \cos\theta_2 & 0 & \sin\theta_2 & a_2 \sin\theta_2 \\ \sin\theta_1 \sin\theta_2 & \cos\theta_1 & -\sin\theta_1 \cos\theta_2 & -a_2 \sin\theta_1 \cos\theta_2 - a_1 \sin\theta_1 \\ 0 & 0 & 0 & 1 \end{pmatrix}.$$

Sample Constellation

The robot is stretched straight up (vertically).

$$\theta_1 = -\frac{\Pi}{2}, \qquad \theta_2 = 0.$$

$$T_2 = \begin{pmatrix} 0 & -1 & 0 & 0 \\ 1 & 0 & 0 & 0 \\ 0 & 0 & 1 & a_1 + a_2 \\ 0 & 0 & 0 & 1 \end{pmatrix}.$$

7.2.6 Example: Mechanism with a Translational and a Rotational Joint

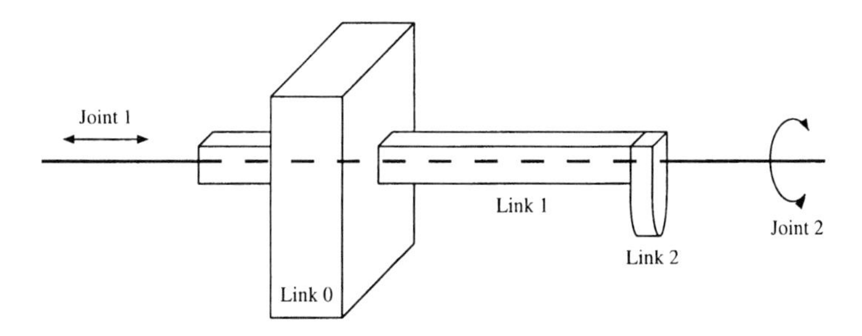

Figure 7.11: Sketch of the mechanism

Denavit-Hartenberg Parameters

Link	a	α	d	θ
1	0	0	d_1	$\frac{\Pi}{2}$
2	0	0	0	θ_2

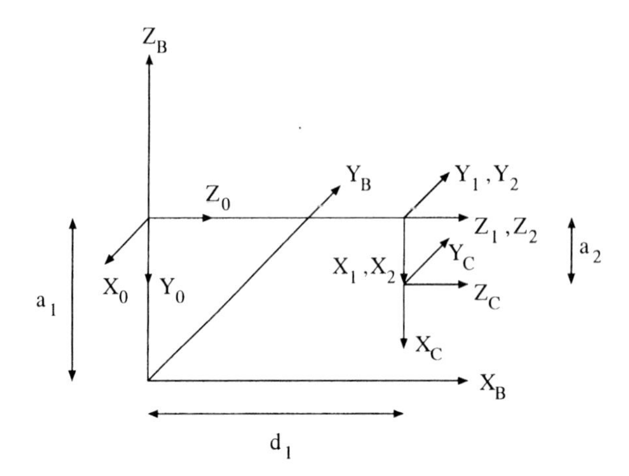

Figure 7.12: Coordinate frames for the joints of the mechanism

Note: The robot model could be optimized by rotating coordinate system 0 about the Z_0 axis by $+90^o$. This would lead to the constant θ_1 being 0. However, θ_1 has been kept non-zero for the sake of illustrating the non-trivial case.

Matrices

$$A_1 = \begin{pmatrix} 0 & -1 & 0 & 0 \\ 1 & 0 & 0 & 0 \\ 0 & 0 & 1 & d_1 \\ 0 & 0 & 0 & 1 \end{pmatrix}.$$

$$A_2 = \begin{pmatrix} \cos\theta_2 & -\sin\theta_2 & 0 & 0 \\ \sin\theta_2 & \cos\theta_2 & 0 & 0 \\ 0 & 0 & 1 & 0 \\ 0 & 0 & 0 & 1 \end{pmatrix}.$$

$$B = \begin{pmatrix} 0 & 0 & 1 & 0 \\ -1 & 0 & 0 & 0 \\ 0 & -1 & 0 & a_1 \\ 0 & 0 & 0 & 1 \end{pmatrix}.$$

$$C = \begin{pmatrix} 1 & 0 & 0 & a_2 \\ 0 & 1 & 0 & 0 \\ 0 & 0 & 1 & 0 \\ 0 & 0 & 0 & 1 \end{pmatrix}.$$

Since $T_2 = B\ A_1\ A_2\ C$,

$$T_2 = \begin{pmatrix} 0 & 0 & 1 & d_1 \\ \sin\theta_2 & \cos\theta_2 & 0 & a_2 \sin\theta_2 \\ -\cos\theta_2 & \sin\theta_2 & 0 & a_1 - a_2\cos\theta_2 \\ 0 & 0 & 0 & 1 \end{pmatrix}.$$

7.2.7 Example: Kinematics of the Stanford Manipulator

The following example has been taken from [Paul, 1981].

The so-called *Stanford manipulator* (see Fig. 7.13) is a robot with 6 axes, one of them being translational. This layout makes it an excellent tool for pick-and-place operations. In Fig. 7.13 the coordinate frames have already been assigned to the links and joints of the robot. Thus, the set of Denavit-Hartenberg parameters can be derived using the procedure described above. One feasible set of Denavit-Hartenberg parameters of the Stanford Manipulator is listed in Tab. 7.1.

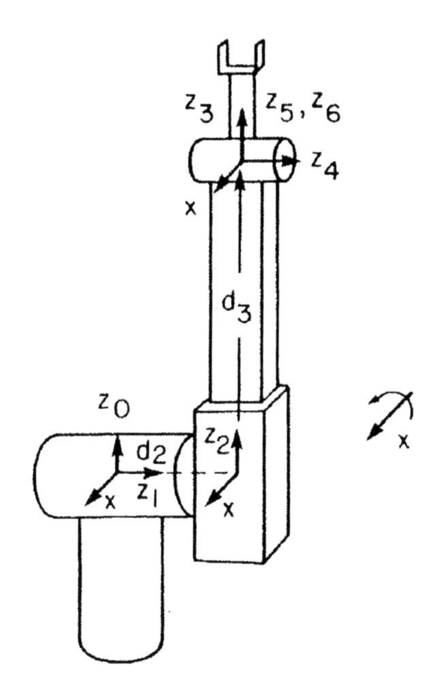

Figure 7.13: Coordinate frames for the Stanford manipulator

These Denavit-Hartenberg parameters yield the following A matrices for the Stanford manipulator:

$$A_1 = \begin{bmatrix} \cos\theta_1 & 0 & -\sin\theta_1 & 0 \\ \sin\theta_1 & 0 & \cos\theta_1 & 0 \\ 0 & -1 & 0 & 0 \\ 0 & 0 & 0 & 1 \end{bmatrix}$$

Link	a	α	d	θ
1	0	$-90°$	0	θ_1
2	0	$90°$	d_2	θ_2
3	0	$0°$	d_3	0
4	0	$-90°$	0	θ_4
5	0	$90°$	0	θ_5
6	0	$0°$	0	θ_6

Table 7.1: Link parameters for the Stanford manipulator

$$A_2 = \begin{bmatrix} \cos\theta_2 & 0 & \sin\theta_2 & 0 \\ \sin\theta_2 & 0 & -\cos\theta_2 & 0 \\ 0 & 1 & 0 & d_2 \\ 0 & 0 & 0 & 1 \end{bmatrix}$$

$$A_3 = \begin{bmatrix} 1 & 0 & 0 & 0 \\ 0 & 1 & 0 & 0 \\ 0 & 0 & 1 & d_3 \\ 0 & 0 & 0 & 1 \end{bmatrix}$$

$$A_4 = \begin{bmatrix} \cos\theta_4 & 0 & -\sin\theta_4 & 0 \\ \sin\theta_4 & 0 & \cos\theta_4 & 0 \\ 0 & -1 & 0 & 0 \\ 0 & 0 & 0 & 1 \end{bmatrix}$$

$$A_5 = \begin{bmatrix} \cos\theta_5 & 0 & \sin\theta_5 & 0 \\ \sin\theta_5 & 0 & -\cos\theta_5 & 0 \\ 0 & 1 & 0 & 0 \\ 0 & 0 & 0 & 1 \end{bmatrix}$$

$$A_6 = \begin{bmatrix} \cos\theta_6 & -\sin\theta_6 & 0 & 0 \\ \sin\theta_6 & \cos\theta_6 & 0 & 0 \\ 0 & 0 & 1 & 0 \\ 0 & 0 & 0 & 1 \end{bmatrix}$$

The products of the A matrices and the further calculations can be read in [Paul, 1981].

7.3 The Inverse Kinematics of a Mechanism

7.3.1 The Problem of Inverse Kinematics

In the previous section we have presented a method for obtaining the location of the end effector frame of a mechanism with respect to some base frame, provided that the Denavit-

Hartenberg parameters are available. In this section we are concerned with the inverse problem. Suppose that the robot-specific D-H parameters a_n, α_n have been fixed. Then, in practical applications one is frequently concerned with solving the

Problem of Inverse (or Backward) Kinematics

Given: T_m
(i. e. the location of the end effector with respect to the base frame), and
$a_1, \cdots, a_m, \alpha_1, \cdots, \alpha_m$.

Find: $d_1, \ldots, d_m$,
$\theta_1, \cdots, \theta_m$
(the so-called joint vector).

Expressed within our mathematical framework, solving the inverse kinematics of a particular robot means solving the following system of equations:

$$T_m = \begin{bmatrix} a_{11} & a_{12} & a_{13} & a_{14} \\ a_{21} & a_{22} & a_{23} & a_{24} \\ a_{31} & a_{32} & a_{33} & a_{34} \\ 0 & 0 & 0 & 1 \end{bmatrix},$$

where each a_{ij} is a (known) function of the D-H parameters a_n, α_n, d_n, θ_n for $1 \leq n \leq m$. Hence, one has to solve 12 independent, non-linear, and transcendental equations in m variables.

7.3.2 Approaches to Solving Kinematic Equations

Up to now, no general, efficient algebraic method is known for solving the inverse kinematics of an arbitrary open kinematic chain (consisting of 6 joints)! Handling *singularities* and *multiple solutions*, see Fig. 7.14 [Craig, 1986], constitutes the major difficulties:

- **Singularities** are positions of the mechanism (particularly the gripper), where infinitesimal changes of the gripper frame lead to (theoretical) discontinuities in the joint space.
- **Multiple solutions** are the general case in the inverse kinematics problem, meaning that the same gripper position and orientation can often be reached by different sets of joint values. Knowing all possible solutions is quite desirable in problems like, e. g., collision avoidance or path optimization.

Numerical methods for solving kinematic equations, unfortunately, usually are able to find (at most) one solution, but not all possible solutions! Furthermore, they are – naturally – very sensitive with respect to problems due to singularities.

With the upcoming of program packages for symbol manipulation, a lot of effort has been undertaken in order to arrive at a symbolic solution for the inverse kinematics problems.

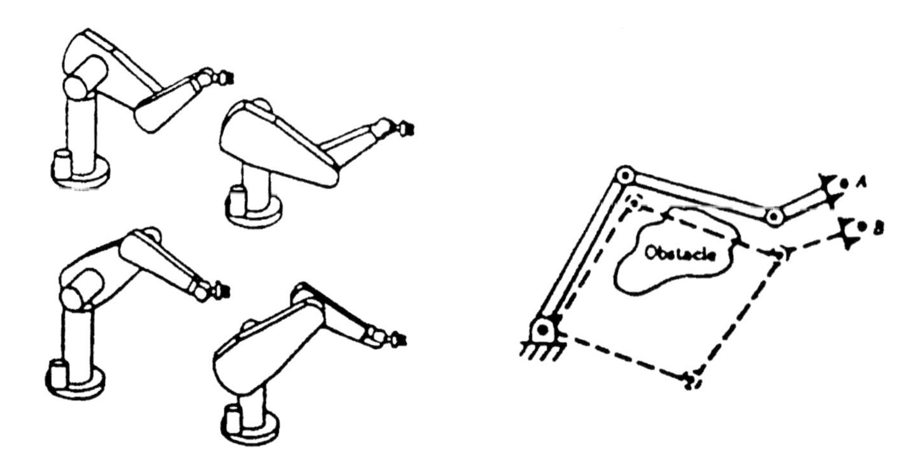

Figure 7.14: All possible joint vectors have to be computed

The advantage of a symbolic ("closed") solution is that the evaluation of expressions is generally faster than a numerical solution and it is definitely more exact. Furthermore, multiple solutions are given by their symbolic equivalents (e. g., equations).

In the following, we will concentrate on a heuristic / algorithmic approach to get a closed solution for the inverse kinematics of open-looped mechanisms with up to 6 degrees of freedom.

7.3.3 An Algorithmic / Heuristic Approach Using A Matrices

In the following, we present the approach given in [Paul, 1981] and realized, e. g., in [Hintenaus, 1987].

Suppose that $T_m = A_1 \cdots A_m$. This equation can be transformed obtaining the equivalent equations

$$\begin{aligned}
T_m &= A_1 \cdots A_m, \\
A_1^{-1} \cdot T_m &= A_2 \cdots A_m, \\
&\vdots \\
A_{i-1}^{-1} \cdots A_1^{-1} \cdot T_m &= A_i \cdots A_m, \\
&\vdots \\
A_{m-1}^{-1} \cdots A_1^{-1} \cdot T_m &= A_m.
\end{aligned}$$

In this set of equations, products of trigonometric functions are replaced using the identities

$$\sin\alpha \sin\beta = \frac{\cos(\beta - \alpha) - \cos(\alpha + \beta)}{2},$$

$$\cos\alpha \cos\beta = \frac{\cos(\beta - \alpha) + \cos(\alpha + \beta)}{2},$$

$$\sin\alpha\cos\beta = \frac{\sin(\alpha+\beta) - \sin(\beta-\alpha)}{2}.$$

Then, the proposed algorithm keeps looking for five characteristic types of equations that can be easily solved:

1. A pair of equations $a\sin\alpha = b \wedge c\cos\alpha = d$, where $a \neq 0$ and $d \neq 0$. This yields

$$\alpha = \arctan\frac{bc}{ad}.$$

Note: For the implementation of the function, **atan2** is to be preferred, i. e. $\alpha =$ atan2 (bc, ad). **atan2** considers the sign of the arguments correctly and it is also defined for $ad = 0$ (if $bc \neq 0$).

2. An equation $a\sin\alpha + b\cos\alpha = c$, where $a \neq 0$ or $b \neq 0$. This yields

$$\alpha = \arctan\left(\frac{-b}{a}\right) - \arctan\left(\frac{c}{\pm\sqrt{a^2+b^2-c^2}}\right).$$

3. A pair of equations $a\sin\alpha + b\sin\beta = c \wedge a\cos\alpha + b\cos\beta = d$, where $a \neq 0$ and $b \neq 0$. This yields

$$\alpha - \beta = \arccos\left(\frac{c^2+d^2-a^2-b^2}{2ab}\right).$$

4. A pair of equations $a\sin\alpha + b\cos\beta = c \wedge a\cos\alpha + b\sin\beta = d$, where $a \neq 0$ and $b \neq 0$. This yields

$$\alpha + \beta = \arcsin\left(\frac{c^2+d^2-a^2-b^2}{2ab}\right).$$

5. A linear equation in a prismatic joint parameter. This yields the straightforward arithmetic expression for the joint parameter.

Hence, trying to solve the inverse kinematics in a symbolic way can be performed as stated in the following procedure:

Procedure InverseKinematics ($\downarrow A_1, \ldots, A_m, T_m, \uparrow$ joint values)

begin
  $i := 1$;
  repeat
    Compute $A_{i-1}^{-1} \cdot A_{i-2}^{-1} \cdots A_1^{-1} \cdot T_m = A_i \cdot A_{i+1} \cdots A_m$, and collect the equations in the entries of the left-hand side and the right-hand side.
    repeat
      Search for a match with one of the patterns described above.
      if there is a match with one of the patterns described above **then**
        Compute solution for the joint parameter.
      endif
    until no more match is found

```
        i := i + 1;
    until a solution for all joints has been obtained or
            no more equations can be generated
    if a solution for all joints has been obtained then
        Report 'Success'
    else
        Report 'Failed'
    endif
end   { InverseKinematics }
```

One should observe that this method does not guarantee to find a symbolic solution in all cases! However, in case a solution can be computed, it is guaranteed that all numerical values for possible joint values can be obtained. If "failed" is being reported, the partial results gained still can give some hints on the structure of the problem (critical parameters, parameters that should be included / excluded in the description of the inverse kinematics, etc.).

For a comprehensive example on how to apply this strategy, we refer to [Paul, 1981].

7.4 Modeling the Kinematics of NC Machines

7.4.1 Typical NC Machine Kinematics

Standard types of NC machines that can be computer programmed for milling, drilling, and turning operations are characterized by kinematics that solely consist of open kinematic chains (i. e. all mechanisms except parallel ones). In the following we present a modeling scheme that allows the efficient implementation of a kinematics module particularly suited for NC machines. NC machines for special tasks, like profile turning or tapping, are not considered.

7.4.1.1 General Kinematic Concept

NC machines can be considered as a special class of mechanisms. By analyzing typical machines one observes that the kinematic structure of the majority of NC machines is very similar and offers several ways of making the computation of the forward kinematics and the inverse kinematics very efficient compared with an arbitrary mechanism, e. g., the kinematic structure of a robot.

When structuring the components of an NC machine, a scheme can be developed consisting of a common stationary base (the *body* of the NC machine) and two (or three) kinematic chains mounted to this base (Fig. 7.15).

The first kinematic chain is formed by the carriage of the machine (*carriage mechanism*) that is affixed to the machine body with one end. The workpiece is mounted onto the opposite end of this chain. If the workpiece is mounted directly to the body (i. e. no carriage is available

B Base
$\mathbf{M_C}$ Carriage Mechanism
$\mathbf{M_T}$ Transmission Mechanism
$\mathbf{CF_C}$ Center Frame for the Workpiece
$\mathbf{CF_T}$ Center Frame for the Tool
□ Translational Joint
○ Rotational Joint

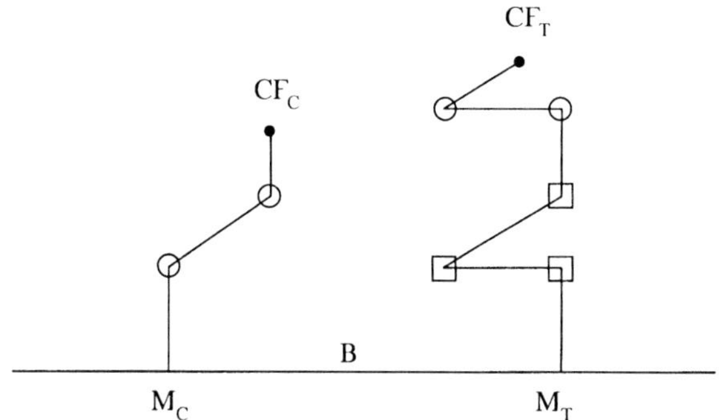

Figure 7.15: Kinematic scheme of a typical NC machine

on that specific machine) the kinematic chain is said to be of "zero-length". For carriage mechanisms, the following kinematic types can occur:

- Type **T**[**T**][2]: The carriage mechanism has one or two translational joints ("sliding carriage").
- Type **R**[**R**]: The carriage mechanism has one or two rotational joints ("tipping carriage").

The second kinematic chain forms the transmission of the machine (*transmission mechanism*). If the fixture of the tool also constitutes a kinematic chain, its kinematic description is adjoined to the description of the transmission (this can be done dynamically during the machining process, i. e. when a tool is mounted). In case of 4D turning (twin-lathe turning) two such kinematic chains are described by the two respective turning transmissions. The first three links of the transmission mechanism (the first two for 2D milling and turning) are translational joints. This is due to stability and positioning quality. If the fixture of the tool does comprise a kinematic chain and in case of 5D milling, one or two rotational joints are adjoined to the translational joints. The different types of machines that fit into this model are listed in Chapter 7.4.1.2.

7.4.1.2 Characteristic Types of NC Machines

In Table 7.2, we present the types of NC machines that have been analyzed and can be programmed by means of a computer. Their kinematic description is given together with specific restrictions for certain types. In the table, we identify joints with their axes. R ∈

[2]Throughout this section, **T** stands for a translational joint, **R** for a rotational one. A joint in squared brackets [ ] indicates that the joint is optional.

$\{T_1, T_2\}$ reads that the axis of joint R corresponds either to the axis of joint T_1 or to the axis of joint T_2.

Process Type	*Carriage Mech.*	*Transmission Mech.*	*Restrictions*
3D Milling		$M_T = \mathbf{TTT}$	
3D Milling with Kinematic Tool		$M_T = \mathbf{TTTR[R]}$	
3D Milling with Tipping Carriage	$M_C = \mathbf{R[R]}$	$M_T = \mathbf{TTT}$	
3D Milling with Sliding Carriage	$M_C = \mathbf{T[T]}$	$M_T = \mathbf{T[T]}$	
3D Milling with Kinematic Tool and Sliding Carriage	$M_C = \mathbf{T[T]}$	$M_T = \mathbf{T[T]R[R]}$	
5D Milling		$M_T = \mathbf{TTTRR}$	
2D Turning	$M_C = \mathbf{R}$	$M_T = \mathbf{T_1T_2}$	$R \in \{T_1, T_2\}$
4D Turning	$M_C = \mathbf{R}$	$M_{T_1} = \mathbf{T_{11}T_{12}}$ $M_{T_2} = \mathbf{T_{21}T_{22}}$	$R \in \{T_{11}, T_{12}\}$, $R \in \{T_{21}, T_{22}\}$
Combined Milling / Turning ("MillTurn")	$M_C = \mathbf{R_1}$	$M_T = \mathbf{T_1T_2T_3R_4[R_5]}$	$R_1, R_4, R_5 \in$ $\in \{T_1, T_2, T_3\}$

Table 7.2: Characteristic types of NC machines

7.4.2 Integrating Machine Geometry and Machine Kinematics

The connection between the geometric description and the kinematic description is done via the link names that are assigned to the Denavit-Hartenberg parameters of the joints. Each link has to be specified as one branch in the layer tree describing the geometry of the NC machine. The order has to be the same in the layer tree and in the kinematic chain. Joints correspond to certain edges in the layer tree.

See Fig. 7.16 for a correct layer tree description of a 3D milling machine with a 2D tipping carriage that uses the kinematics of the example in Chapter 7.5.2. (The tool, the workpiece, and the fixtures are not described. Extra layer trees are necessary for their description.)

7.5 Example: Kinematics of a 5D Milling Machine

7.5.1 Forward Kinematics Description

For describing the forward kinematics of the kinematic chains of NC machines we use the Denavit-Hartenberg parameters in the way described in [Paul, 1981]. The workspace of the NC machine may be restricted by specific limits for each of the joints. Thus, for every joint an interval may be specified indicating the range of allowed joint values. For translational joints, the extrusion range of the joint is specified; for rotational joints the range of the joint angle is specified. If no interval is specified, the joint is supposed to be able to move without any restriction.

For machining steps with impact on the tool the velocity profile of the single joints is resolved already by the postprocessor. It is therefore available in the control code directly as geometry and kinematic information. In order to achieve a realistic simulation of the rapid feed, a trapezoidal velocity profile is specified for each of the joints by the parameters *acceleration*,

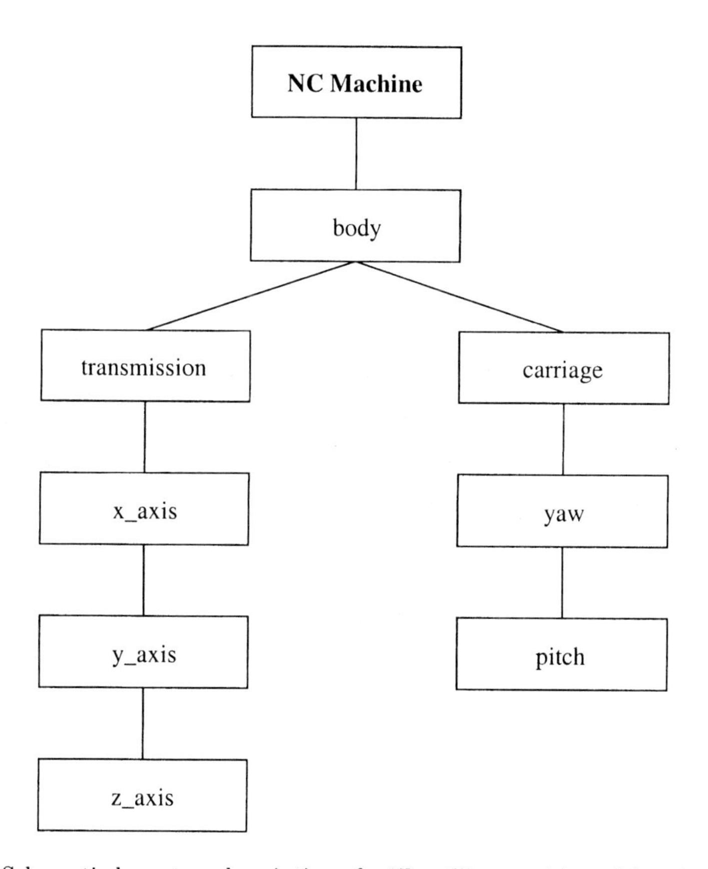

Figure 7.16: Schematic layer tree description of a 3D milling machine with a tipping carriage

maximal velocity, and *deceleration*. The joints are then moved in a synchronous way from the start position to the end position, according to these parameters.

The layer tree is used for the description of the geometry of the mechanisms involved in the manufacturing simulation. The kinematics of these mechanisms is described in a separate *kinematics description file*.

In order to allow easy data exchange, the kinematics description file is – like the layer tree – in ASCII format. In this file we describe the forward and inverse kinematics for every open kinematic chain. The structure of the file is the following:

```
kinematicDescriptionList:
    /* empty */
  | kinematicDescription  kinematicDescriptionList

kinematicDescription:
    forwardKin
    inverseKin

forwardKin:
    frame
    jointParameterList
    frame

jointParameterList:
    jointParameter
  | jointParameter jointParameterList

jointParameter:
    layerElementName '=' denavitHartenbergParameters
                         jointRange
                         velocityProfile


denavitHartenbergParameters:
    'T' '<' a alpha theta d '>'
  | 'R' '<' a alpha theta d '>'

jointRange:
    '[' ']'
  | '[' min max ']'

velocityProfile:
    '(' ')'
  | '(' acceleration maxVelocity deceleration ')'

frame:
    '[' ']'
  | '[' f1 ... f12 ']'
```

Remarks:

- The `denavitHartenbergParameters` are specified in the following way: The description begins with a name that identifies the element in the layer tree where the kinematic parameters belong to. More precisely, the Denavit-Hartenberg parameters describe the relationship of the named element and its father node.

  The next character entry specifies the type of the joint: `R` defines a rotational joint and `T` defines a translational joint. After the joint type, the Denavit-Hartenberg parameters ($< a\ \alpha\ \theta\ d >$) themselves have to be specified. For a translational joint, the specified value for parameter d (being a variable by the definition of a translational joint) defines the "home" position of the joint; for a rotational joint, the parameter θ fulfills the same purpose.

- In order to guarantee that the specified set of Denavit-Hartenberg parameters describe a valid *kinematic chain*, the following condition must be fulfilled for every joint description: If j_1 and j_2 are the names of two succeeding joints, then j_1 and j_2 must be valid names in the corresponding layer tree and j_1 must be the father of j_2.

- The `jointRange` defines the interval of valid joint values. If the interval is omitted, no range checking will be performed.

- The `velocityProfile` defines the velocity behavior of the joint at rapid feed. We assume a trapezoidal velocity profile. The first parameter specifies the acceleration (in m/s^2 for translational joints and rad/s^2 for rotational joints). The second parameter specifies the maximum velocity (in m/s, rad/s respectively). The third parameter specifies the non-negative deceleration.

- The frame preceding the joint parameters is a fixed homogeneous transformation that defines the relationship between the reference coordinate system and the base coordinate system of the Denavit-Hartenberg parameters. We call this frame *base frame* (*BF*).

  The frame succeeding the joint parameters defines the *center frame* (*CF*) for the workpiece or the tool (often called *gripper frame* when speaking of industrial robots only. This frame should be oriented as stated in [Paul, 1981]. Thus, for tools that are symmetric with respect to an axis the z-axis of the tool frame (the so-called *approach vector*) must be coincident with the rotation axis of the tool.

  An empty frame denotes the unit frame.

7.5.2 Forward Kinematics Values

The Denavit-Hartenberg parameters of transmission and the tipping carriage of the NC machine shown in Fig. 3.3 are summarized in the following two tables:

Transmission				
Link	a	α	d	θ
1	0	$\frac{\pi}{2}$	t_1	0
2	0	$\frac{\pi}{2}$	t_2	$\frac{\pi}{2}$
3	0	0	t_3	$\frac{\pi}{2}$

$$BF_T = \begin{bmatrix} 0 & 0 & 1 & 0 \\ 0 & -1 & 0 & 0 \\ 1 & 0 & 0 & 0 \\ 0 & 0 & 0 & 1 \end{bmatrix}, \quad CF_T = \begin{bmatrix} 1 & 0 & 0 & 0 \\ 0 & 1 & 0 & 0 \\ 0 & 0 & 1 & 0 \\ 0 & 0 & 0 & 1 \end{bmatrix}$$

Tipping Carriage				
Link	a	α	d	θ
1	a_1	$-\frac{\pi}{2}$	0	r_1
2	a_2	0	0	r_2

$$BF_C = \begin{bmatrix} 0 & 0 & 1 & 0 \\ 1 & 0 & 0 & 0 \\ 0 & 1 & 0 & 0 \\ 0 & 0 & 0 & 1 \end{bmatrix}, \quad CF_C = \begin{bmatrix} 0 & 0 & 1 & 0 \\ 0 & -1 & 0 & 0 \\ 1 & 0 & 0 & 0 \\ 0 & 0 & 0 & 1 \end{bmatrix}$$

If we assume the link lengths $a_1 = 20.0$ and $a_2 = 10.0$ the forward kinematics can be described as follows:

```
[ 0  0  1
  0 -1  0
  1  0  0
  0  0  0 ]

x_axis = T <0.0 1.5708 0.0    0.0>  [ 0.0    500.0] (2.0 5.0 2.0)
y_axis = T <0.0 1.5708 1.5708 0.0>  [ 0.0    500.0] (2.0 5.0 2.0)
z_axis = T <0.0 0.0    0.0    0.0>  [-100.0 100.0] (2.0 5.0 2.0)

[ ]  -- unit frame

-- The inverse kinematics description of the transmission
-- mechanism has to be inserted here.


[ 0  0  1
  1  0  0
  0  1  0
  0  0  0 ]

yaw   = R <20.0 -1.5708 0.0 0.0> [-0.7854 0.7854] (2.0 3.5 2.0)
pitch = R <10.0  0.0    0.0 0.0> [-0.7854 0.7854] (2.0 3.5 2.0)

[ 0  0  1
  0 -1  0
  1  0  0
```

```
  0  0  0 ]

-- The inverse kinematics description of the tipping carriage
-- has to be inserted here.
```

7.5.3 Inverse Kinematics Description

For describing the inverse kinematics of the mechanisms of an NC machine an ASCII file description in C-style syntax is used. The formulas defined therein are interpreted during the simulation run. Every joint parameter is defined by a symbolic expression (a set of algebraic or trigonometric equations) together with conditions when this expression is applicable. In this way, the selection of the correct expression in the case of non-unique solutions can be controlled by the user. By defining a file with the kinematics description a whole kinematics library for NC machines can be established. If desired, this library can be extended by the user without the necessity to adapt other components of the system.

In older systems, the inverse kinematics are described in a subroutine that has to be compiled and linked to the existing executable file. This procedure is very cumbersome and time-consuming.

To increase flexibility, we offer the possibility to describe the inverse kinematics of every NC machine on a separate ASCII file and to dynamically load and interpret the inverse kinematics of the machine to be simulated (see Ch. 11.8.2 for details on our system). The inverse kinematics of an open kinematic chain has to be added to the kinematics description file just after the parameters of the corresponding forward kinematics.

The inverse kinematics can be described in C-like style. The following grammar shows the details of this description language:

```
inverseKin: /* empty */
          | '{' paralist statementlist '}'

paralist:   /* empty */
          | inpara paralist
          | outpara paralist

inpara:   IN ':' varlist ';'

outpara: OUT ':' varlist ';'

varlist:    var
          | var ',' varlist


statementlist: /* empty */
             | statement statementlist

statement:
```

```
               ';'
             | '{' statementlist '}'
             | if_stat
             | while_stat
             | ret_stat   ';'
             | exp        ';'

if_stat:       IF '(' exp ')' statement else_stat



else_stat: /* empty */
             | ELSE statement

while_stat: WHILE '(' exp ')' statement


ret_stat: RETURN  exp

exp:         num
            | var
            | var '=' exp
            | fnct  '(' exp ')'
            | fnct2 '(' exp ',' exp ')'
            | exp '<'  exp
            | exp '<=' exp
            | exp '>'  exp
            | exp '>=' exp
            | exp '==' exp
            | exp '!=' exp
            | exp '||' exp
            | exp '&&' exp
            | exp '+'  exp
            | exp '-'  exp
            | exp '*'  exp
            | exp '/'  exp
            | '-' exp
            | '!' exp
            | exp '^' exp  /* exponentiation (same as 'pow') */
            | '(' exp ')'

fnct:      'sin'
          | 'cos'
          | 'tan'
          | 'asin'
          | 'acos'
          | 'atan'
          | 'log'
          | 'log10'
```

```
       | 'sqrt'
       | 'fabs'
       | 'exp'
       | 'floor'
       | 'ceil'

fnct2:   'atan2'
       | 'pow'
```

7.5.4 Interface for the Inverse Kinematics Description

- **in-parameters:** Some or all of the parameters of T are passed to the inverse kinematics module. Depending on the number of declared input parameters these parameters are initialized with values of the matrix tool frame

$$T = \begin{bmatrix} n_x & n_y & n_z & 0 \\ o_x & o_y & o_z & 0 \\ a_x & a_y & a_z & 0 \\ p_x & p_y & p_z & 1 \end{bmatrix}$$

  in the following way:

  - **12 parameters:** $n_x, n_y, n_z, o_x, o_y, o_z, a_x, a_y, a_z, p_x, p_y, p_z$ are assigned to the input parameters 1 to 12 (in this order).
  - **6 parameters:** $a_x, a_y, a_z, p_x, p_y, p_z$ are assigned to the input parameters 1 to 6 (in this order).
  - **3 parameters:** p_x, p_y, p_z are assigned to the input parameters 1 to 3 (in this order).

  Any other number of input parameters is invalid.

- **out-parameters:** From these input parameters the joint values are computed according to the inverse kinematics description. The variables that contain the result for the joint values have to be declared as `out`-variables. The order of these variables defines how the joint values are mapped to joints, i. e. the ith `out`-variable is mapped to the ith joint in the kinematic chain.

  The number of `out`-variables has to be equal to the number of joints in the forward kinematics description.

- **return-value:** A non-zero `return`-value (i. e. TRUE) signals the calling procedure that a solution to the inverse kinematics problem was found. If `return`-value is zero (i. e. FALSE) the `out`-variables may be undefined.

7.5.5 Inverse Kinematics Values

Since the transmission mechanism of our sample NC machine of Fig. 7.15 consists only of translational joints, the inverse kinematics is of course very easy:

$$t_1 = p_x,$$
$$t_2 = p_y,$$
$$t_3 = p_z.$$

The inverse kinematics of the tipping carriage computes to:

$$r_1 = \begin{cases} asin(\frac{a_x}{\sqrt{1-a_y{}^2}}) & \text{if } |a_y| < 1 \\ acos(n_x) & \text{otherwise} \end{cases},$$

$$r_2 = -asin(a_y).$$

This formula can be translated into the inverse kinematics description language in a straightforward way:

- **Inverse kinematics description for the transmission mechanism:**

```
{
   in:  px, py, pz;
   out: t1, t2, t3;

   TRUE = 1;

   t1 = px;
   t2 = py;
   t3 = pz;

   return TRUE;
}
```

- **Inverse kinematics description for the tipping carriage:**

```
{
   in:  nx, ny, nz, ox, oy, oz, ax, ay, az, px, py, pz;
   out: r1, r2;

   TRUE = 1; FALSE = 0;
   EPS = 0.001;

   if (az < 0)   /* unreachable approach vector */
      return FALSE;

   if (fabs(ay) < 1-EPS)
      r1 = asin(ax/sqrt(1-ay^2));
   else
      r1 = acos(nx);
```

```
    r2 = -asin(ay);

    return TRUE;
}
```

Chapter 8

Data Accuracy and Consistency

This chapter states and classifies the problems arising from inaccurate and / or inconsistent data. Based upon this classification, strategies for solving and avoiding these problems are presented.

8.1 Reasons for Inconsistent Data

Most objects in the real world belong to a continuous domain, but data structures and algorithms can only use discrete representations and computations. Therefore one can distinguish two error reasons,

1. **external errors**, arising from an inaccurate mapping of real-world data,
2. **internal errors**, due to discrete computation and representation.

In many cases the simulation of a continuous domain with floating point numbers leads to acceptable results, even without consideration of inexactness. However, there are many problems where this approach will fail for certain inputs.

Consequently, one can denote the following two problem levels:

1. **quantitative problems**, i. e. inexact results,
2. **qualitative problems**, i. e. wrong results.

In many cases, quantitative problems are the reason for qualitative problems at a later stage.

Representations of problems that behave in such a way are, e. g.:

- **Input-sensitive algorithms:** Due to the strong connection between input data for an algorithm and the exactness of the output one can distinguish
  - *well-conditioned input* and
  - *ill-conditioned input*

for a given algorithm.

As an example, the angle symmetry for two vectors in 2D can be computed by adding the corresponding unit vectors (Fig. 8.1).

Figure 8.1: Angle symmetry with well-conditioned input

In Fig. 8.1 the input (vectors $\vec{a}$ and $\vec{b}$) is well-conditioned, but an input like Fig. 8.2 leads to problems, because the angle between $\vec{a}$ and $\vec{b}$ is close to 180°.

Figure 8.2: Angle symmetry with ill-conditioned input

- **Transformation of representations:** A transformation of geometry that can be represented by integers sometimes leads to a geometry that can not be represented by integers in a strict sense. For instance, consider a square with edge length a, for which the edge points can be represented by integers (Fig. 8.3). Rotate this square by 30°. The resulting square can neither be represented using integers nor using rationals, because position p_1 is $(\frac{\sqrt{3}}{2}a, \frac{1}{2}a)$.

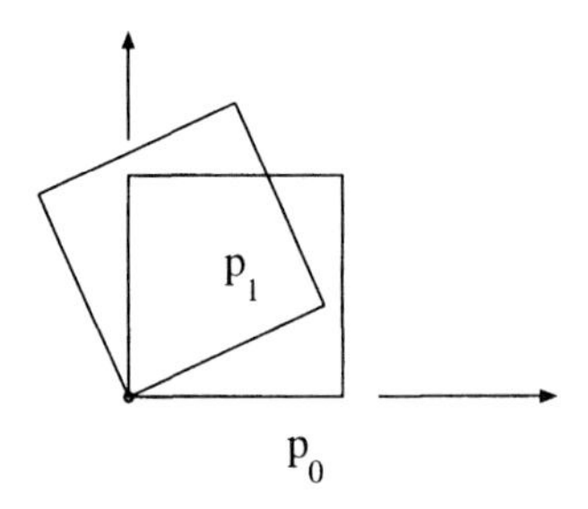

Figure 8.3: Rotation of a square

In the next section, we will classify such problems in order to structure them and derive solutions.

8.2 Classification of Accuracy and Consistency Problems

The main problem to be solved by geometric algorithms is to decide, whether objects intersect each other or have common elements. Therefore we need reliable algorithms to compute the topological data of geometric objects.

In such algorithms, geometric problems generally (see [Spur *et al.*, 1988a]) arise when:

- the distance or the angle between two objects is close to the computational exactness of the algorithm or the data structure, or
- coincidence of objects is given.

Main sources for errors are:

- unknown exact values (e. g., every measuring is inexact),
- representation errors (internal representations of numbers have limits for the smallest and the largest number that can be represented),
- round-off errors,
- coincidence of geometric objects:
  - partial coincidence, e. g., the non bordered geometry is the same,
  - totally coincidence, e. g., partially coincident and the border is the same,
- (wrong) termination conditions of numerical methods.

The *reliability* of a geometric modeler[1] is determined by the degree of mastering these error sources. Factors characterizing the reliability are depicted in Fig. 8.4.

These factors already indicate possible solutions for achieving modeling reliability:

- higher internal precision,
- better algorithms,
- database of geometric problems,
- interval arithmetic,
- better data structures:
  - separation of given data and computed data,
  - feature modeling.

In the next section, these strategies will be analyzed in detail.

[1]Following [Spur *et al.*, 1988a], we define the *reliability* of a geometric modeler as the ability to guarantee correct functionality despite of the failure of single components.

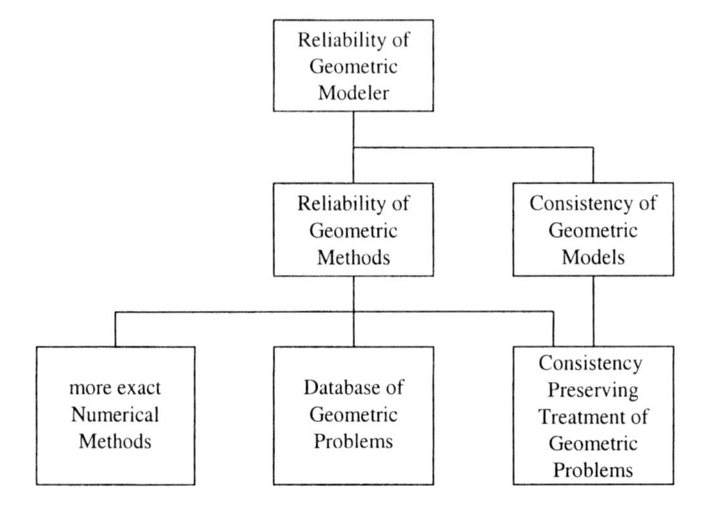

Figure 8.4: Factors for reliability of geometric modeler

8.3 Solutions

8.3.1 Higher Internal Precision

The use of higher internal precision, e. g., floating point numbers with double or triple precision, for geometric algorithms narrows the range for *ill-conditioned input*, but will not solve the problem of failing of an algorithm in general.

The following two examples show that qualitative errors can never be avoided using the strategy of increasing the internal precision.

Example 8.3.1.1 *Incidence Asymmetry*

Given: 4 Lines: L_1, L_2, L_3, L_4, where
i_1 is the intersection of L_1 and L_2,
i_2 is the intersection of L_3 and L_4.

Decide: Are i_1 and i_2 equal?

One way of solving this problem is:

1. Compute i_1.
2. Prove, whether i_1 is on L_3 and L_4.

or, alternatively:

1. Compute i_2.

2. Prove, whether i_2 is on line L_1 and L_2.

Using exact numbers makes the two solutions equivalent, but using floating point numbers with any precision forces us to assume a distance in which incidence is valid. Such a distance is shown in Fig. 8.5 as a dashed line specifying the incidence area for the given line intersections.

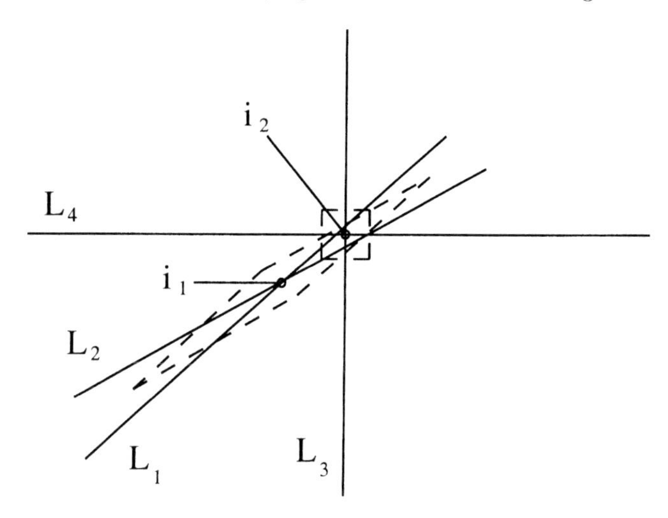

Figure 8.5: Incidence area and asymmetry

One can see that computing i_1 and checking with proximity to L_3 and L_4 leads to the result that L_1, L_2, L_3, and L_4 have no common intersection point, whereas computing i_2 and checking with proximity to L_1 and L_2 – in contradiction – indicates that a common intersection point does not exist.

Example 8.3.1.2 – ***The Pentagon Problem***

Several geometric problems can only be solved step by step. Each transformation performs a loss of precision. However, many operations have no inverse transformation even when using exact computation. This can be demonstrated, e. g., by the so-called *pentagon problem* (Fig. 8.6; [Hoffmann, 1989]):

On a pentagon one defines two functions, `In` and `Out`. `In` constructs an inscribing pentagon using the five diagonals of the original pentagon, `Out` extends the five sides of the original pentagon to their outer intersections, thus defining a circumscribing pentagon. In many cases the `Out` operation will be the inverse operation to `In`. But for the pentagon in Fig. 8.7 the first `In` leads to a quadragon for which the `Out` operation is undefined. When using non-exact computation, this problem does not only occur for this single special case, but also for all pentagons of approximately the same shape. This increases the number of ill-conditioned problems dramatically.

Nevertheless, increasing internal precision is an inexpensive (and therefore frequently applied) method to reduce the number of ill-conditioned inputs and therefore has its right of existence.

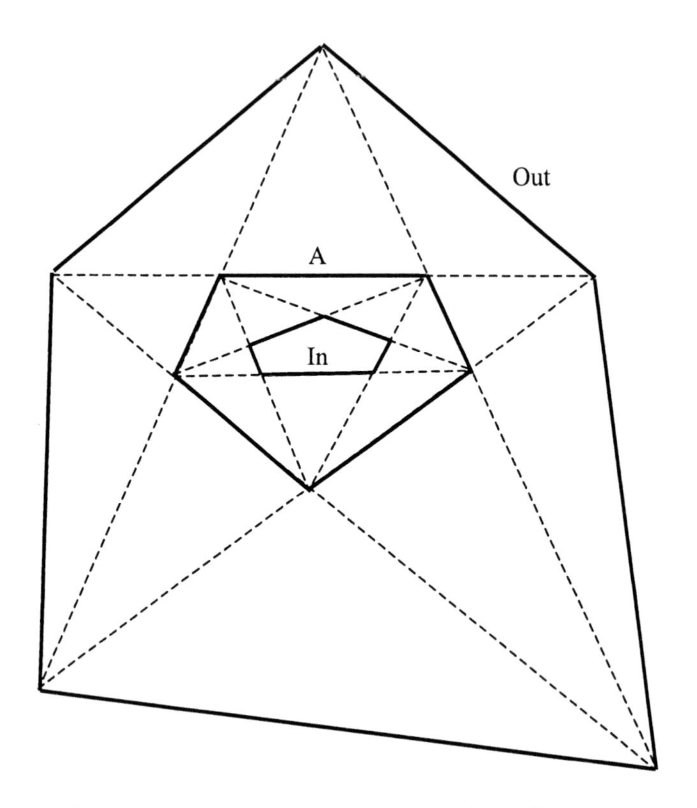

Figure 8.6: Pentagon problem: `In` and `Out` functions

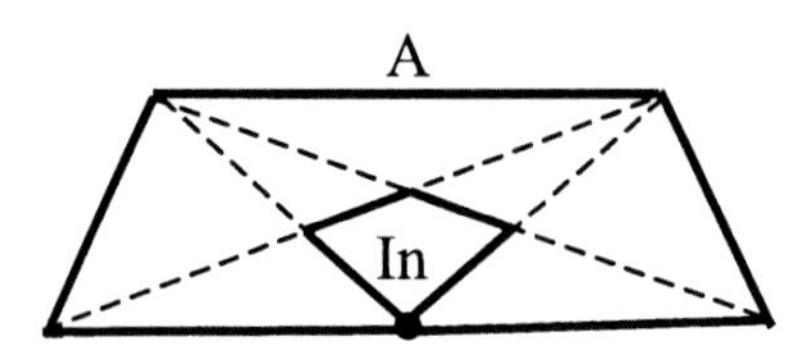

Figure 8.7: Ill-conditioned pentagon

8.3.2 Better Algorithms

In general, one struggles for better geometric algorithms in order to preserve well-conditioned geometries throughout tasks like transformations or evaluations. We want to demonstrate this idea on the problem of determining, whether a point is contained in an area, i. e.

Given: Point p and area F.

Decide: Is point p contained in area F ?

In the literature, this problem is often called *point-in-polygon test*, see, e. g., [Lutz, 1988].

Two algorithms are presented for performing this test, the *vector method* and the *differential geometric method*.

Example 8.3.2.1 – ***Vector Method***

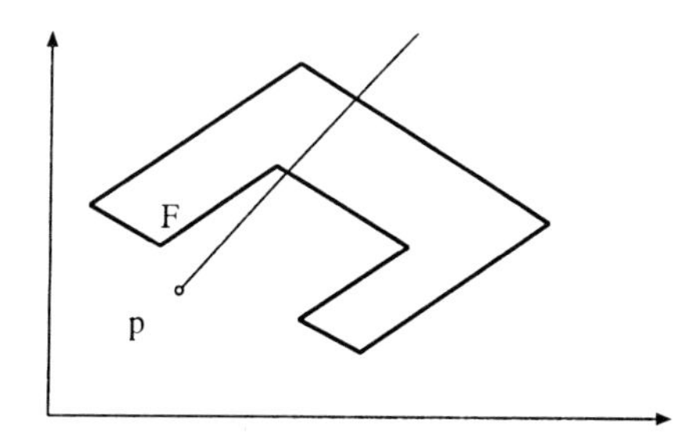

Figure 8.8: Vector method for the point-in-polygon test

The algorithmic idea is the following (Fig. 8.8):

1. Construct an arbitrary line through point p.
2. Count the intersections between the borderline of the area.
3. Resolutions:
   (a) If the number of intersections is odd, p is in area F.
   (b) If the number of intersections is even, p is not in area F.

This strategy works well if the intersections are not coincident with vertices of the area. For such cases the algorithm may return the wrong result.

This geometric instability led to the development of higher sophisticated algorithms, like the *differential geometric method.*

Example 8.3.2.2 – ***Differential Geometric Method***

The algorithmic idea of this approach is to consider the point environment. This allows us to prove simpler geometric structures instead of the given geometry. The main steps of this method are:

1. Compute a vector v into an arbitrary direction from p.
2. If there is a intersection between v and the borderline, consider the nearest intersection point as reference point p_r.
3. Consider the tangent vectors in the environment of p_r.
4. Resolutions:
   (a) If the distance between p and p_r equals zero, p is on the area.
   (b) If there is no intersection, p is not on the area.
   (c) If p is on the left side of a tangent vector, p is on the area, else p is not on the area due to the direction of the tangent vector defined as shown in Fig. 8.9.

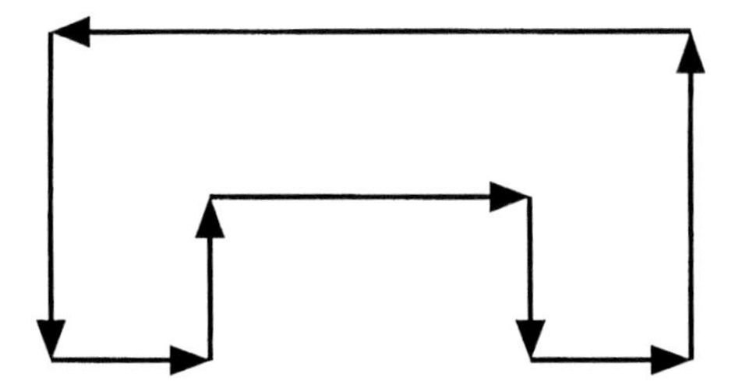

Figure 8.9: Definition for the direction of a tangent vector

   We need three additional vectors for the next step of the algorithm (Fig. 8.10).
   (a) r_0 pointing from p_r to p,
   (b) r_a pointing into the tangent direction,
   (c) r_e pointing into reverse direction of the tangent.
5. Compute the cross product $w_a \times w_0$ of the difference vectors $w_a = r_a - r_e$ and $w_0 = r_0 - r_e$.
   (a) If $w_a \times w_0$ is equal to the zero vector, the result is undefined.
   (b) If the direction of $w_a \times w_0$ points into the same direction as the normal vector on the area, p is on the area F else p is not on the area F.

A general problem of algorithm design is that certain results that should already have been determined and treated as a special case (or consequences of such results) might appear again

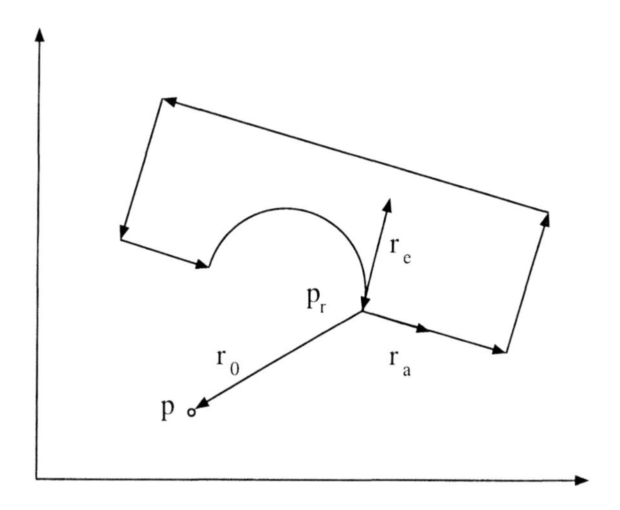

Figure 8.10: Direction vectors in p_r

at a later stage. At this stage, however, that cases are no longer expected to be possible and, consequently, not handled correctly or not handled at all.

Summarizing, struggling for "better" algorithms can mean algorithms of higher stability, but also algorithms that can be applied to a wider input range.

8.3.3 Problem Database

Consistency in logic means that the formal system is free of contradictions. We can apply this definition to geometric structures in the following way: Transformations that are basically consistent have to be consistent in a geometric modeler, i. e. transformed consistent models must, again, constitute consistent models. Therefore we can say that geometric modelers are libraries for reliable geometric transformations.

Known geometric problems can be added to a geometric modeler by means of rules that characterize a specific situation or a class of situations ("*analysis*") and specify a response for these cases ("*action*"). Although the enumeration of error cases makes no sense from a theoretical point of view (one will always leave out infinitely many problems), practical experience shows that in most cases all but very few of the errors can be analyzed and taken into account within reasonable effort.

An example for such a rule is the *separation rule* (see [Spur *et al.*, 1988b]) for preserving the correct topological structure of a boundary-represented (B-Rep, see Chapter 6.5) object. For defining this rule, we need the following

Definition 8.3.1 (Bound) The *boundary* of a three-dimensional solid is the union of the bounding planes.

A *point* is the intersection of two bounding elements of a plane, An *edge* is the intersection of two bounding planes of a body, A *plane* bounds exactly one body.

Using this hierarchical definition of relations between geometric elements, we define the *separation rule*:

Definition 8.3.2 (Separation Rule) *If a topological condition for the number of relations of a geometric element to the elements of the (hierarchically) next higher level is violated, the involved element must be duplicated. The hierarchically higher element must be split into two (or more) separate elements.*

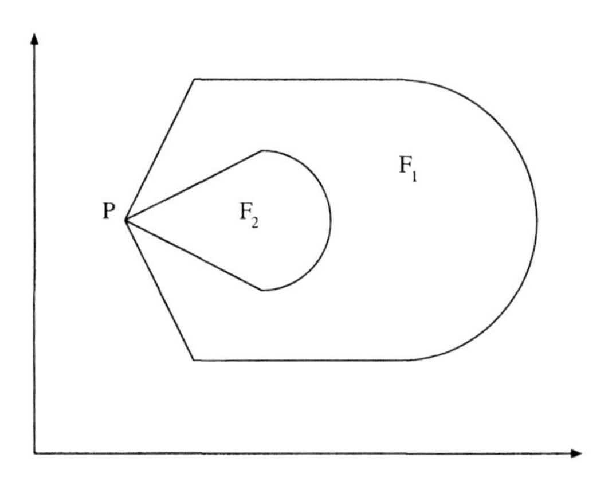

Figure 8.11: Intersection of two areas, F_1 and F_2

In the example shown in Fig. 8.11 we have to decide, whether

- the areas F_1 and F_2 intersect each other or
- the area F_2 is contained in F_1.

The intersection of F_1 and F_2 yields a point (being of lower dimension than the areas) and, thus, violates the conditions for bounded solids. Therefore we have to examine point P and its environment ϵ. For any environment ϵ for P, all points in F_2 that are not equal to P lie in F_1. Using the theory of limits, the properties of ϵ are assigned to P ("continuous extension"). Consequently, F_2 lies in F_1.

In [Hoffmann, 1989], this strategy is called *normalizing* the input data. There, the postulate is that no two vertices are allowed to be closer than some tolerance ϵ and that, likewise, no vertex can be closer to an edge than ϵ. Two functions are necessary for altering the input adequately, *edge cracking* and *vertex shifting*.

Summarizing, the following more general rules can be derived from this example:

- Keep the data consistent in any geometric situation.
- Boundary elements get the properties of their environment if consistency is not given otherwise.
- Apply the separation rule in such a way that geometric adaptations are minimal (this makes loops of adaptations impossible).

8.3.4 Interval Arithmetic

A good representation for inexact data is an interval describing the lower and upper bound for a value. For such an interval we can define the basic operations $(+, -, \cdot, /)$ and the main properties of other operations.

The main advantage of interval arithmetic is that the result is an interval containing the exact value. In this way we know, on each calculation step, the exactness of the results and therefore we can decide whether the results are good enough or not.

A classical example for the application of interval arithmetic is the *point incidence check*, i. e. checking two points for equality using a defined point environment. Two different formulas for defining point incidence for two points $x = (x_1, x_2, x_3)$, $y = (y_1, y_2, y_3)$ in three-dimensional space can be found quite frequently:

1. **Sphere method**:
$$\sqrt{\sum_{i=1}^{3}(x_i - y_i)^2} < \epsilon,$$
2. **Box method**:
$$\bigwedge_{1 \leq i \leq 3} | x_i - y_i | < \epsilon.$$

The "exactness factor" ϵ can be chosen in three different ways:

1. a fixed value throughout the whole system,
2. an appropriate value for each algorithm,
3. an appropriate value for each number.

The third alternative, constituting the so-called *interval arithmetic* in the classical sense will be discussed further in the following. It should be obvious, however, that also the other two alternatives follow the same theory.

8.3.4.1 Definitions

We use the following conventions:

- a, b are variables in the set of the real numbers $\mathcal{R}$.
- i, m, n are variables in the set of the natural numbers $\mathcal{N}$.

Definition 8.3.3 (Interval I in $\mathcal{R}$) *I, $I := [a, b]$, is an interval iff I is a subset of the real numbers R, $I \subseteq \mathcal{R}$, where a is the lower bound and b is the upper bound, i. e. $[a, b] := \{x | a \leq x \leq b\}$.*

By defining an interval as a set, the set operations $\in, \cap, \cup, \subset, \subseteq$ are quite clear. In the following we write $\mathcal{I}$ when we mean the universal set of intervals.

Definition 8.3.4 (Width([a,b])) *The width of an interval is given by the distance between a and b. Because of Def. 8.3.3, where $a \leq b$, we can define $Width([a,b]) := b - a$.*

Definition 8.3.5 (Ordering <) *Let $[a,b],[c,d] \in \mathcal{I}$. Then $[a,b] < [c,d]$ iff $b < c$.*

Definition 8.3.6 (Equality =) *Two intervals, $[a,b],[c,d] \in \mathcal{I}$, are equal, $[a,b] = [c,d]$, iff $a = c, b = d$.*

Definition 8.3.7 (Point Interval) *An interval I, $I := [a,b]$, is a point interval iff $a = b$. In the following we write a in the meaning of [a,a].*

Definition 8.3.8 (Box) *$x \in \mathcal{I}^n$ is an n-dimensional box.*

Definition 8.3.9 (Sub-box) *Let $X, Y \in \mathcal{I}^n$, $X := (X_1, \ldots, X_n)$, $Y := (Y_1, \ldots, Y_n)$. Then X is sub-box of Y iff for all i, $i \in \mathcal{N}$, $X_i \subseteq Y_i$.*

Definition 8.3.10 (Interval Operation) *Let $A, B \in \mathcal{I}$. $\circ$ is an interval operation iff $A \circ B = \{a \circ b | a \in A, b \in B\}$.*

Definition 8.3.11 (Inclusion Function) *F, $F : \mathcal{I}^n \to \mathcal{I}^m$, is an inclusion function for f, $f : \mathcal{R}^n \to \mathcal{R}^m$, iff for a given point interval x, $f(x) \in F(x)$.*

Definition 8.3.12 (Inclusion Interval Extension) *An inclusion function F, $F : \mathcal{I}^n \to \mathcal{I}^m$, is called an inclusion interval extension for f, $f : \mathcal{R}^n \to \mathcal{R}^m$, iff for a given point interval x, $f(x) = F(x)$.*

Definition 8.3.13 (Inclusion Monotony) *F, $F : \mathcal{I}^n \to \mathcal{I}^m$, is an inclusion monotonous function iff for each sub-box X of box Y, F(X) is sub-box of F(Y).*

8.3.4.2 Basic Operations on Intervals

According to Def. 8.3.10 on interval operations, we define the basic operations $\{+, -, ., /\}$ as follows:

Let $X := [a,b], Y := [c,d] \in \mathcal{I}$.

$$\begin{array}{lllllll}
X & + & Y & = & [a,b] & + & [c,d] = [a+c, b+d], \\
X & - & Y & = & [a,b] & - & [c,d] = [a-d, b-c], \\
X & . & Y & = & [a,b] & . & [c,d] = [min(ac,bc,ad,bd), max(ac,bc,ad,bd)], \\
X & / & Y & = & [a,b] & / & [c,d] = [a,b].[1/c, 1/d] \text{ iff } 0 \notin [c,d].
\end{array}$$

8.3.4.3 Inverse Elements On Intervals

Lemma 8.3.1 (Inverse Elements) *There are no inverse elements for the basic operations* $\{+, -, \cdot, /\}$ *in* $\mathcal{I}$, *except for point intervals.*

Proof: We demonstrate the proof only on the operation $+$. We start with the point interval 0. $0 = [0, 0]$ as in Def. 8.3.7 for point interval. $[0, 0] = [a - a, b - b]$ as given by the rules for real numbers. $[a - a, b - b] \stackrel{?}{=} [a, b] + [-a, -b]$ as defined in section 8.3.4.2, If we except point intervals, we assume that $a < b$ and $-a < -b$ (Def. 8.3.3), but this is a contradiction. □

8.3.4.4 Machine Interval Arithmetic

Machine interval arithmetic is the approximation of interval arithmetic on computer systems, due to the fact of limited precision of numbers in computers. Therefore, for an interval I we define its machine representation (*machine interval*) M as an interval $M = [a_m, b_m]$, where a_m and b_m are values that can be represented in the computer, of smallest width such that $I \subseteq M$.

8.3.4.5 Logical Problems with Intervals

We have defined the equality (Def. 8.3.6) for intervals. However, there is a gap between equality and inequality, as we can see in Fig. 8.12.

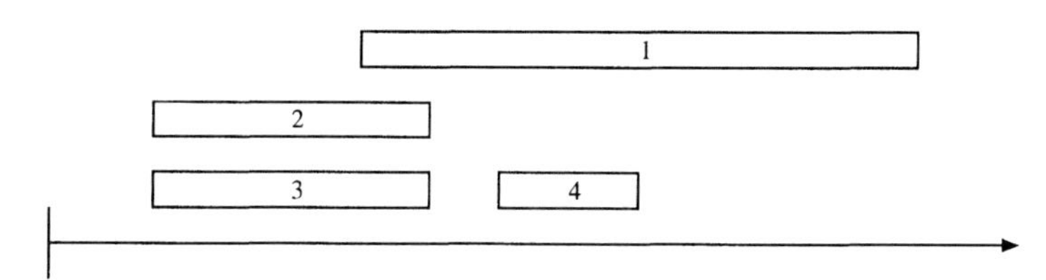

Figure 8.12: Ordering and equality

In this example we can decide that $interval_2 \neq interval_4$ and $interval_2 = interval_3$.

But we cannot say anything about the relation between $interval_1$ and $interval_2$. Therefore we have three possible answers for given intervals $A, B \in \mathcal{I}$:

1. EQUAL, where $A = B$,
2. NOT EQUAL, where $A \neq B$ and
3. UNDEFINED, else.

For the smaller relation "<", the situation is similar.

Type	Data 1	Data 2	Action
1	given	given	error message and abort
2	given	computed	change data 2
3	computed	given	change data 1
4	computed	computed	change data 1 and 2

Table 8.1: Actions when determining two inconsistent pieces of data

8.3.4.6 Reasoning on Intervals

Reasoning methods on intervals are determined by their underlying logic. Main logic schemes that are used are:

- **Binary logic:** In *binary logic*, we can only decide whether an interval has a given property or not. We can write this decision in the set notation as $a \in M$ or $a \notin M$ with a as an element with certain properties and M as the set where all the elements with these properties are contained. We have discussed the insufficiency of such a binary logic in conjunction with intervals in Chapter 8.3.4.5.

- **Possibilistic logic:** *Possibilistic logic* is another approach to the membership of elements. The so-called *on belief function* [Hajek and Harmanec, 1992] defines the membership of elements as $m(a) = p_M$, where a is used as in binary logic and p_M is the possibility m for the membership of a in a given set M. In binary logic the definition of m is $m : U \to \{0, 1\}$, which differs from the possibilistic definition $m : U \to [0, 1]$, where U denotes the universal set.

- **Fuzzy logic:** The main difference between *fuzzy logic* and possibilistic logic is that in fuzzy logic $m(a)$ denotes a function that defines the degree of membership for an element a. In this way, instead of defining the possibility of membership, the degree is specified how similar an element is to a given concept M.

8.3.5 Better Data Structures

8.3.5.1 Separation of Given Data and Computed Data

It is useful to store the input data from the user separate, because we have to assume correctness for such inputs. Computation should never change these (correct) inputs without warning. Detected inconsistency should always perform changes on the computed data. Later computed data should be changed first to guarantee consistency. The appropriate actions when determining two inconsistent pieces of data are listed in Table 8.1.

A consistency check must be performed after each geometric computation. However, consistency-preserving actions as in Table 8.1 that follow a situation of type 2, 3 or 4 sometimes lead to type 1 situations, thus making a consistency-preserving continuation impossible. In case a continuation is possible, "corrected" (= adapted) data should be marked appropriately.

A typical case, where consistency-preserving actions are frequently necessary, is the task of filleting, where parts of a geometric contour have to be fitted in order to preserve properties of the whole contour. An example is a circular arc inscribed into two lines in such a way that the directions of the two lines should coincide with the direction of the tangents at the start point and the end point of the arc (Fig. 8.13).

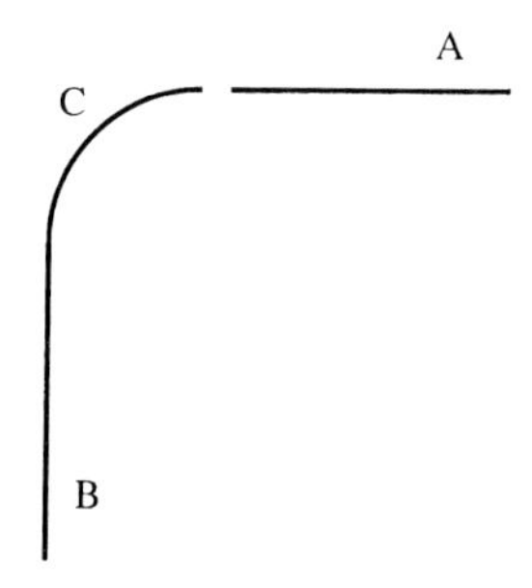

Figure 8.13: Tangential inscription of a circular arc

This problem can be described in the following way:

Given: two lines A and B.

Find: a tangential circular arc C between A and B without changing the edge points of A and B.

It is obvious that the problem is over-determined by this description, because either

- two points, a radius and a given side for the center point, or
- two tangents, a radius and free edge points of the tangents,

already constitute sufficient geometric information. For this problem and similar ones, the adaptation of one or more of the geometric elements is necessary in order to achieve consistency. This is a rather intuitive and therefore very difficult task. Furthermore, one adaptation may lead to inconsistencies at other parts of the contour. Algorithms for such tasks are consequently highly problem-specific.

8.3.5.2 Feature Modeling

Particularly due to hybrid modeling, but sometimes specially for consistency purposes, it gets increasingly popular to store the information of a geometric object in a redundant way. As an example, for a square, a reference point plus side length and angle can be stored additionally to its four corner points, see Fig. 8.14.

Storing additional (redundant) geometric information enables more comprehensive geometric consistency checks. The redundancy can additionally be used for tracking and "correcting" consistency errors, similar to error-correcting codes. By introducing information like lines being *perpendicular* onto each other, or lines being *parallel*, additional reasoning (e. g., normal $\circ$ normal $\equiv$ parallel) can be performed. Such a geometric description being comprised of

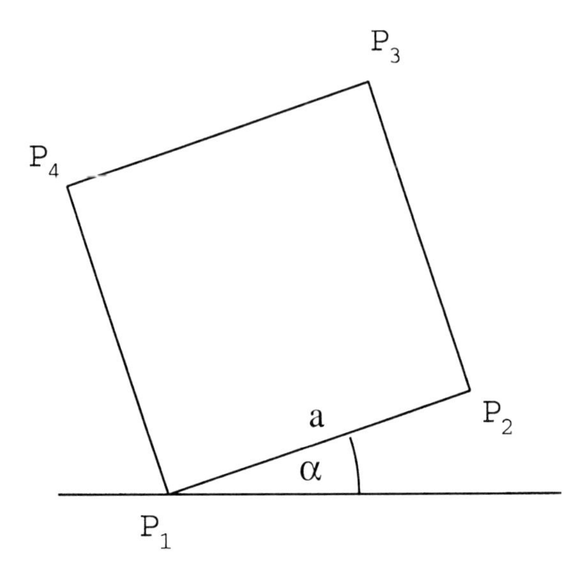

Figure 8.14: Redundant description of a square

data together with rules is frequently called *feature*. Certain aspects of this so-called *feature modeling* technique are closely related to *artificial intelligence* and *knowledge engineering* topics.

8.4 Example: The GEM System for Configuration Modeling

We present an engineering framework that is aimed for modeling and configuration of modularly designed mechanical products. The enormous expansion of such products is implied by the modularization of modern manufacturing resulting in a more manageable and economic way of production: Prefabricated modules and flexibly adaptable module frames are specified, adapted, and assembled. Our system supports all these production steps by computer, enabling an ideal combination of mass production of modules composed to individually adapted products satisfying the customer's specific needs (see [Mayr, 1997a] for details on our system).

8.4.1 Motivation: Modularization of Manufacturing

Mechanical products are increasingly being designed and modeled in a modular way. This is implied by the modularization of modern manufacturing concepts in order to manufacture complex systems in a manageable and economic way: Prefabricated modules and flexibly adaptable module frames are selected, customized, and assembled, each step ideally supported by computer. The modularization process defined this way enables an ideal combination of mass production of modules with the composition of individually adapted products.

Typical products that follow this modular concept are prefabricated houses, furniture systems like, e. g., kitchens or living room components, winter gardens, presentation booths, or modularly designed engineering systems, like car engine / drive-shaft combinations and climatic cabinets. A characteristic feature of all these product classes is that even in companies where the subcontracting of partners supplying prefabricated parts has been very rare, the number of prefabricated components that are manufactured externally (so-called "COTS", i. e. commercial, off-the-shelf products) and are then integrated into the product. This is mainly due to the increasing specialization within the production of certain sub-components with substantial technical know-how (like, e. g., heat exchangers or sound absorbers in the field of climatic cabinets).

In addition to the efforts of designing modular components in a more general way in order to increase their fields of application, this leads to immense variety of possible component combinations that fulfill the constraints of a specific product requirements description. Only by means of modern, engineering-style software support – that has to be used during all phases, like design, contracting, production planning and manufacturing – a suitable product can be offered and produced for specific, individual requirements in an efficient and economical way.

8.4.2 Configuration Modeling

A key factor for a successful product generation using this concept is an optimal *configuration* of the (pre-)selected product components prior to their assembly. The related modeling process is called *configuration modeling*. The main characteristics of configuration modeling is the circumstance that a product has to be composed out of components, regarding various constraints (technical, economical, environmental, etc.) and optimization criteria. Furthermore, the set of available components is being changed continuously by, e. g., product innovations or change of subcontractors. The goal of a software system for configuration modeling therefore has to be to keep the set of basic components and primitives as small and well-structured as possible in order to keep the modeling basis consistent, free of redundancy, and maintainable. Additionally, the degree of automation has to be increased as highly as possible in order to take routine work (like adding the right amount of assembly screws, calculating the cumulated assembly time, or dimensioning simpler components appropriately) off the human expert's shoulders.

Thus, the goal of configuration modeling is to support all key stages of module-oriented production:

- **Specification – definition of components**: All pre-designed and pre-defined instances of meta models constitute the class of readily available modules for the specification of the needed components. If new or adapted modules are needed, they can easily and consistently be derived from the best fitting meta model available.
- **Configuration – adaptation of components**: When using pre-fabricated modules and meta models, some parameters and behavior of these components can remain unspecified in order to be fixed during the defining of the topology of the whole system or device (e. g. defining the glass thickness of a winter garden roof depending on the actual panel size and the inclination of the roof, or specifying a suitable sound absorber

for a climatic cabinet depending on the air streams and the ventilators needed).

- **Assembly – combination of components**: Algorithms and rules specified by the system maintainer enable a valid assembly of the defined configuration, like the computation of the correct number of screws needed for assembling two parts, including their optimal placement, or determining a valid assembly sequence.

8.4.3 An Innovative Solution Based Upon Meta Models

8.4.3.1 Engineering Frameworks

According to [Webster's, 1995], *engineering* is "the application of science and mathematics by which the properties of matter and the source of energy in nature are made useful to human beings", and a *framework* is defined as a "structural or skeletal frame or a basic structure".

Due to the necessity of genericity and flexible adaptability, software frameworks heavily rely on the concepts of object-oriented modeling. In the context of software engineering the term "framework" is often used as an alias for *application framework* which is roughly defined as a collection of software units for a specific application domain (see, e. g., [Pomberger and Blaschek, 1996]). According to [Hametner, 1996], the basic characteristics of an object-oriented engineering framework are:

- The data and knowledge representation must be in an object-oriented manner.
- The whole development process of an engineer must be supported.
- It must be possible to extend the framework by application specific units. The generic slots that can be used in this way for customizing the tool are called *hot spots*.

An object-oriented engineering framework must cover the capabilities of engineering databases (see, e. g., [Eigner *et al.*, 1991]), intelligent computer aided design (see, e. g., [Tomiyama and Yoshikawa, 1990]), discrete event specified systems (see, e. g., [Zeigler, 1984]) and application frameworks in order to support the whole engineering process for an arbitrary application domain [Artmann, 1996].

8.4.3.2 The Engineering Framework GEM

The engineering framework GEM (Generic Engineering framework based upon Meta models) constitutes a state-of-the art engineering environment that supports the engineering (i. e. specification, configuration, and assembly) of modularly designed products by means of *meta models*, i. e. flexible module templates that follow the object-oriented paradigm.

Our focus is the interactive, incremental modeling of mechanical products. The GEM engineering framework supports this task by means of automatically filtering out all non-adequate modules at each stage of modeling as well as performing continuous computation of parameter values in order to specify the correct variant and checking the consistency of the product being modeled. This is done by applying algorithmic and declarative rules describing the system dependencies and the system behavior that can be assigned to each component in the modeling tree by the system maintainer. For the implementation of the algorithmic and rule-based knowledge structure of the system, an adaptation of the DEVS scheme [Zeigler, 1984] has been used.

The four main modeling steps when using GEM are shown in Fig. 8.15 – again taking climatic cabinets as the example:

1. A module is specified by selecting it in the taxonomy of components (i. e. a hierarchically structured component tree). If the module is parameterized, its parameters are fixed as far as this is possible without knowing details about the other modules.
2. The module is added to the product ("device"); configuration constraints are checked automatically.
3. The remaining parameters are fixed and abstract sub-parts are made concrete with the knowledge of the location of the module within the device and the other modules already selected. Standards and technical requirements are checked and the assembly is finalized.
4. After completing the whole device in this way, reports for the product are created (e. g., offers, work plans, bills of material) and made available to the corresponding software systems (e. g., offering, production planning, accounting).

8.4.3.3 The Concept of Meta Modeling

In GEM, *meta models* are components that have been defined in order to compose them into products both "as is" and in an adapted way by means of deriving similar or more detailed components from them. Meta models as a most innovative way of modeling are made possible through the benefits of object orientation, the "ready-to-use" features of frameworks as well as dynamic customization features of object-oriented data bases, see Fig. 8.16. The main advantages are:

- **Engineering view onto modeling**: Components can be selected from the hierarchical component tree and be assembled into the product without specifying all the details (that might not be known at this stage), functioning as component patterns until they can be specified in more detail.
- **Hierarchical administration of component data**: The components are available to the user in a hierarchical component tree. The taxonomy that defines that tree can be specified by the user according to the classification structure he is familiar with. Such a tree allows a step-by-step refinement of requested components as well as an efficient search for similar components.
- **Pre-defined basic configurations**: Frequently needed arrangements of components ("sub-products") can be stored again in a hierarchical way ("configuration tree"). When assembling such a sub-product, the parameters internal to the component are fixed automatically. This feature allows the pre-definition of component arrangement for frequently needed product classes, like, e. g., climatic cabinets for restaurants or hospitals.

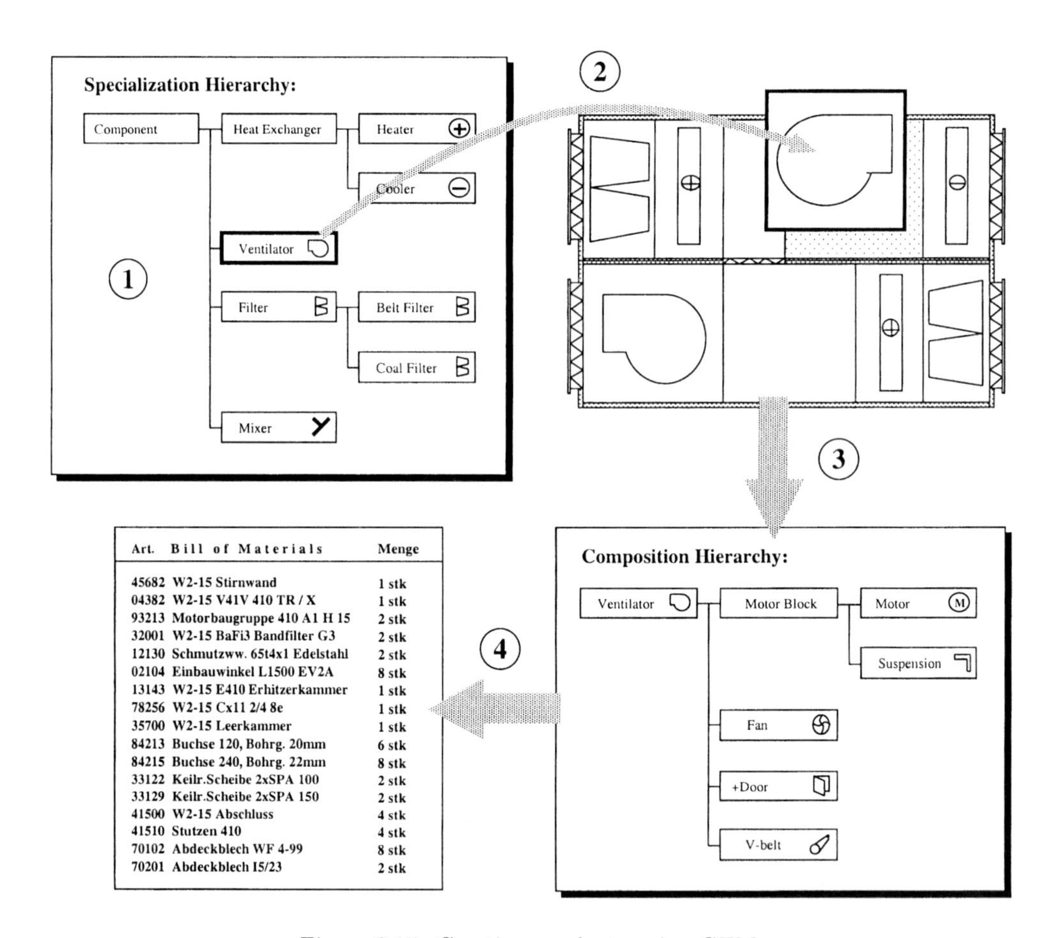

Figure 8.15: Creating products using GEM

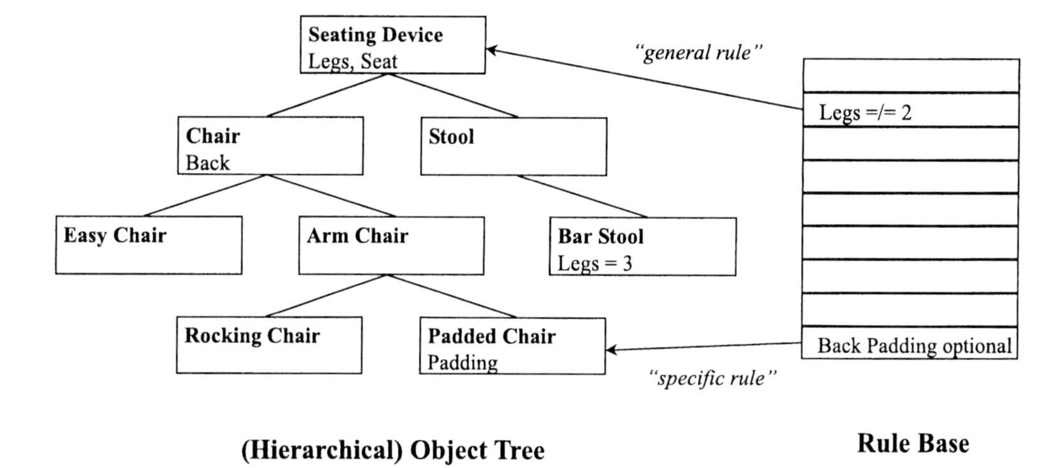

Figure 8.16: The concept of meta models

- **Pre-fabricated templates**: Some component parameters (like color or size) allow a broad variety of possible values. Such parameters can be described by means of component templates, which enables the possibility of choosing any feasible value for a parameter while storing only that components in the component tree whose parameter combinations have already been used. This keeps the component tree small and additionally allows the evaluation, how frequently certain values for parameters are needed. In this way, the degree of detail of the component taxonomy can be adapted dynamically over time to the needs of a manufacturer.

- **Testing of alternative solutions - optimization**: Component patterns and component templates can be refined manually, thus being able to compare the influence of alternative component on the performance of the whole product. Alternatively, according to certain criteria the optimal component can be selected automatically from the sub-tree defined by the component pattern or component template.

- **Management of produced components, products, and projects**: Each created product together with its component structure is added to the configuration tree. Thus, already designed products can be used as a basis and adapted for a product or a subproduct of a new project, enabling a very efficient way of reusing available know-how.

In the next section we illustrate, how algorithms and rules can be adapted by the user in GEM additionally to the static knowledge.

8.4.4 Introducing Dynamics by Means of Algorithms and Rules

As described in the previous section, each component can be assigned attributes and composition relations defining sub-components of the component. Additionally, a freely configurable description of procedural and declarative knowledge is enabled through the features of assigning formulas, algorithms, and rules to each component, see Fig. 8.17.

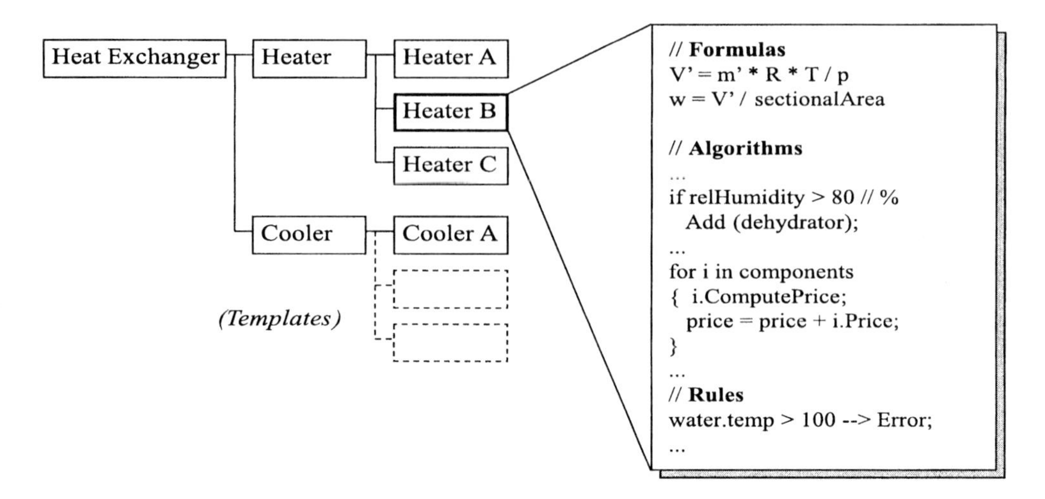

Figure 8.17: The dynamics of the solution: algorithms and rules

Procedural and declarative knowledge can be described using a specially developed description language that basically follows the structure of C and C++, but has been extended by features specifically needed in engineering frameworks. Examples for such features are: the definition of default values and check rules for input parameters; the conditional composition of components; the automatic generation of reports from the actual model.

The definition of formulas, algorithms, and rules within the single components enforces a strict information hiding, leading to a modern programming style, i. e. small, clear procedures "at the right place" [Thomas and Rozenblit, 1995]. Since our programming environment is based on the object-oriented principle, the advantages of genericity ("Every heat exchanger can calculate its efficiency factor."), composability ("Each chamber can use the knowledge defined within the medium 'air'."), and inheritance ("If the general ventilator component comprises a motor, each specific ventilator component comprises a (specific) motor.").

The formulation and adaptation of algorithmic knowledge in GEM is possible without any internal changes or coding at developer level. For details concerning the GEM programming environment we refer to [Hametner, 1996].

8.4.5 Higher-order Meta Models: Meta Modules

GEM comprises a fully functional core system with modules for all desired fields of application, defined at different levels of detail. These modules contain abstract sections that can be adapted by the user according to his specific needs ("meta modules", see Fig. 8.18). Consequently, GEM comprises the module levels:

1. The core system has been implemented in an object-oriented way and comprises the modeling facilities, the user interface components, and the data base interface, an object-oriented extension of SQL.

2. For various fields of application, meta modules have been realized, like, e. g., for con-

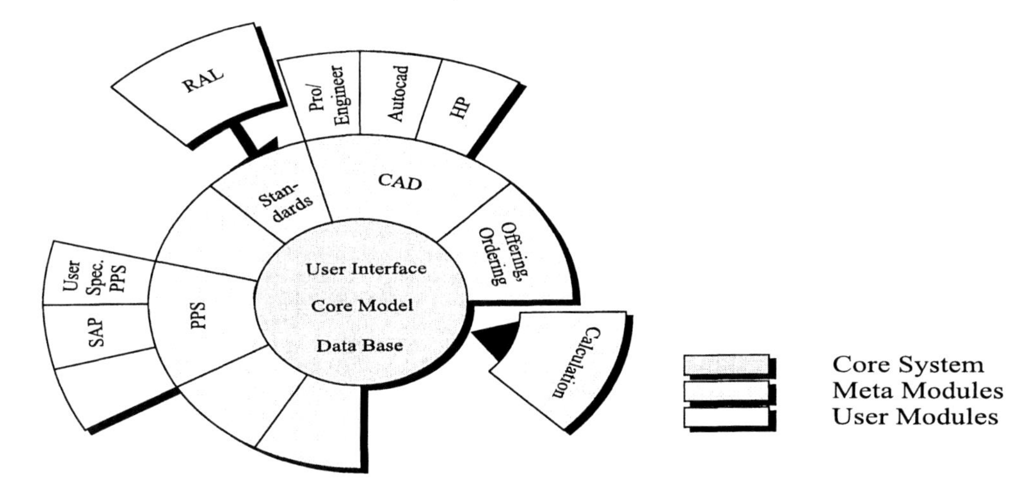

Figure 8.18: Higher-order meta models: meta modules

version of CAD data, contracting, PPS coupling, and work plan generation.

3. Specifically for a certain field of application or company needs, user modules can be added or refined from meta modules using the GEM programming environment. In the frame of our first applications, a customer-specific program library for climatic engineering has been integrated, an HPGL graphic data interface, an interface to a PPS system and a detailed cost accounting interface have been realized.

Due to this three-level design, GEM can be used immediately for a new field of application, but can also be coupled or integrated with a pallet of application-specific modules. The concept of meta modules allows the use of functionalities of the system already immediately after installation. When needed, these functionalities can be adapted appropriately. In this way, the user can customize GEM according to his needs and keep it adaptable without necessity to change any source code. For a more formal approach to describing a multi-level knowledge enhancement, we refer to, e. g., [Rozenblit and Hu, 1992].

8.4.6 Advantages of GEM

The main advantages of our engineering framework GEM are:

- The engineering approach of the production process is reflected in the software.
- Maintenance and extensions can be made by the (authorized) user without changes in the source code.
- Adaptations can be made immediately by the user.
- The framework can be customized for specific application needs.
- The product and process know-how remains with the user and not with the software developer.

Since the component description and the component hierarchy is being designed by an expert of the respective engineering domain from the company, the structure of the software model is a precise image of the structure of the processes, components, and products of the design department and the manufacturing department and, thus, is familiar and manageable for the system maintainer. In this way, the benefits of modern systems engineering and computer science theory are made available to the mechanical engineer and product designer in an environment and a language where such experts feel familiar. Besides of this innovation at the tool level, the main benefits of our engineering framework are products that can be customized to the customer's needs much more precisely already at the stage of requirement analysis and product planning, which increases customer satisfaction as well as reduces planning and production costs.

Our concepts as well as first versions of our product have already been included into a software environment for car engine / drive-shaft test benches and a CAD / CAM / CAP system for climatic cabinets.

Part III

Visualization

Chapter 9

Graphic Algorithms for Visualization

In this chapter, sample algorithms are presented that prepare geometric data in such a way that they can be visualized on modern, window-oriented raster-graphic displays. The importance of time-efficient algorithms is explicitly stressed.

9.1 Window-Viewport Transformation

Each element of a virtual window has to be transformed into pixels on the real raster display ("viewport"), see Fig. 9.1. Since there are – normally – quite a lot of elements to transform and this has to be done quite often a while, these transformations have to be done as efficiently as possible.

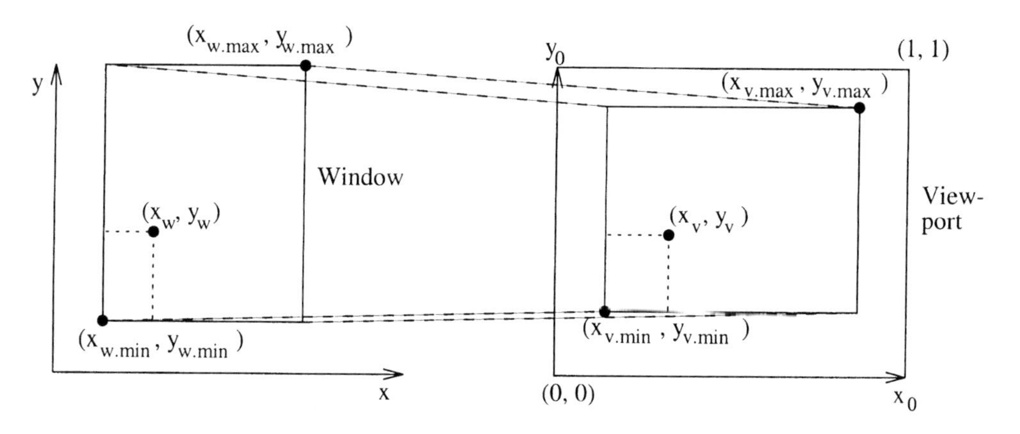

Figure 9.1: Window-viewport transformation

The basic idea is the following: The relative distances within both windows remain unchanged, only scaling and translation is performed. This results in the following ratio:

$$\frac{x_{w.dist}}{x_{w.len}} = \frac{x_w - x_{w.min}}{x_{w.max} - x_{w.min}} = \frac{x_v - x_{v.min}}{x_{v.max} - x_{v.min}} = \frac{x_{v.dist}}{x_{v.len}}$$

$$\frac{y_{w.dist}}{y_{w.len}} = \frac{y_w - y_{w.min}}{y_{w.max} - y_{w.min}} = \frac{y_v - y_{v.min}}{y_{v.max} - y_{v.min}} = \frac{y_{v.dist}}{y_{v.len}}$$

Making x_v and y_v explicit, we obtain

$$x_v = x_{v.min} + \frac{x_{v.max} - x_{v.min}}{x_{w.max} - x_{w.min}}.(x_w - x_{w.min})$$

$$y_v = y_{v.min} + \frac{y_{v.max} - y_{v.min}}{y_{w.max} - y_{w.min}}.(y_w - y_{w.min})$$

These formulae can be further simplified. By substituting

$$s_x = \frac{x_{v.max} - x_{v.min}}{x_{w.max} - x_{w.min}}, s_y = \frac{y_{v.max} - y_{v.min}}{y_{w.max} - y_{w.min}}$$

one obtains

$$x_v = x_{v.min} + s_x(x_w - x_{w.min})$$

$$y_v = y_{v.min} + s_y(y_w - y_{w.min})$$

We call s_x and s_y *scaling factors*; $x_{v.min}$ and $y_{v.min}$ are called *translation factors*.

By applying basic arithmetic, one finally obtains

$$x_v = s_x x_w + t_x$$

$$y_v = s_y y_w + t_y$$

with

$$t_x = x_{v.min} - s_x x_{w.min}$$

$$t_y = y_{v.min} - s_y y_{w.min}$$

The factors s_x, s_y, t_x, and t_y are independent of the image to be transformed. They can be computed in a preprocessing step. For the actual transformation step, only two multiplications and two additions are necessary. This allows fast and easy computations.

9.2 Clipping

Since the visualization area for objects is generally restricted in its size (computer screen, window of a graphic workstation, or sheet of paper on a printer), geometrically transformed objects that exceed these size limits (either the scale factor might be too large or the object itself is of infinite size – like a line) have to be "cut" in an appropriate way in order that only that part of the object is to be drawn that is contained in the visualization area. This cutting process is called *clipping* in graphic data processing.

Usually, (2D) clipping is done in an isothetic rectangle, the *clipping area*. for 3D clipping, an isothetic box defines the restrictions. In the following, we without loss of generality explain methods and algorithms in 2D.

Clipping can be done either in the window prior to the mapping onto the viewport or afterwards, resulting in the same output:

- First clipping of the star in world coordinates to the window boundary, then mapping it to device coordinates, Fig. 9.2:

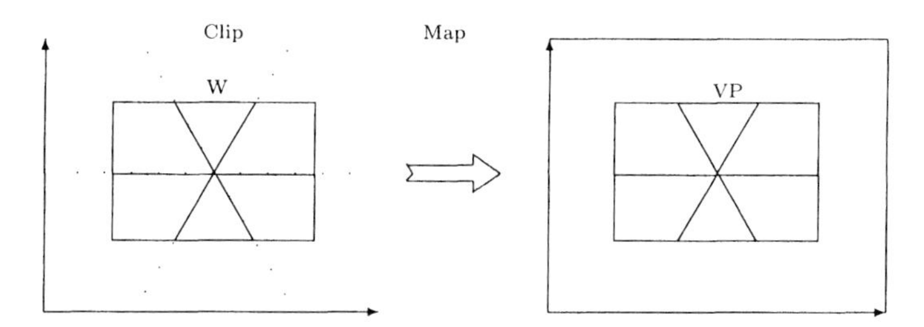

Figure 9.2: The operations `Clip` and `Map`

- First mapping the star to device coordinates, then clipping it to the viewport boundary, Fig. 9.3:

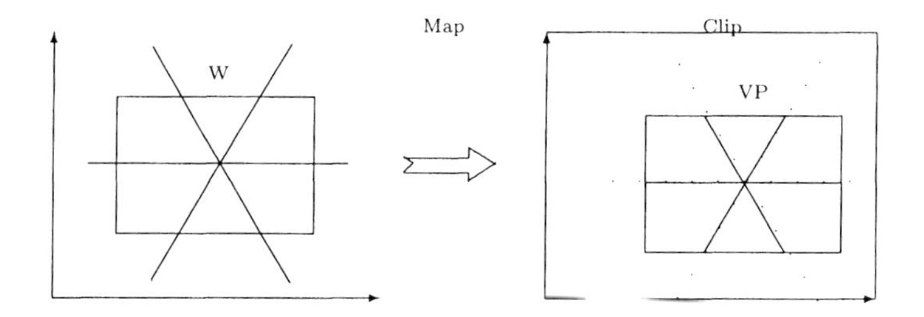

Figure 9.3: `Clip` and `Map` are commutative

If clipping is done in software, it is generally preferred to clip according to the window boundaries prior to the viewport transformation, since this can be done device-independently using world coordinates.

Due to reasons of performance, clipping is increasingly done in hardware. In this case the better approach is to clip – application-independently – on the viewport boundaries using device coordinates.

9.2.1 Clipping of Points

Clipping of points can be done straightforwardly by conditional drawing of a point, since a point $P(x, y)$ is visible only when it lies within the window boundary, i.e. $x_{min} \leq x \leq x_{max}$ and $y_{min} \leq y \leq y_{max}$.

9.2.2 Clipping of Lines

9.2.2.1 Simple Algorithm

A "brute-force" approach to clipping a line would be the computation of its intersection points with all four lines that delimit the window boundary. This would of course work. However, as already mentioned, the efficiency of a clipping algorithm is of major importance for the performance of computing a graphic output. Therefore, this brute-force approach is practically not acceptable since it necessitates four line-line intersection computations per line that is to be drawn.

Consequently, clipping of lines and line segments is done by analyzing the start points and end points of the lines. Algorithms pursuing this idea are presented in the next sections. Fig. 9.4 depicts some of the possible cases of line clipping.

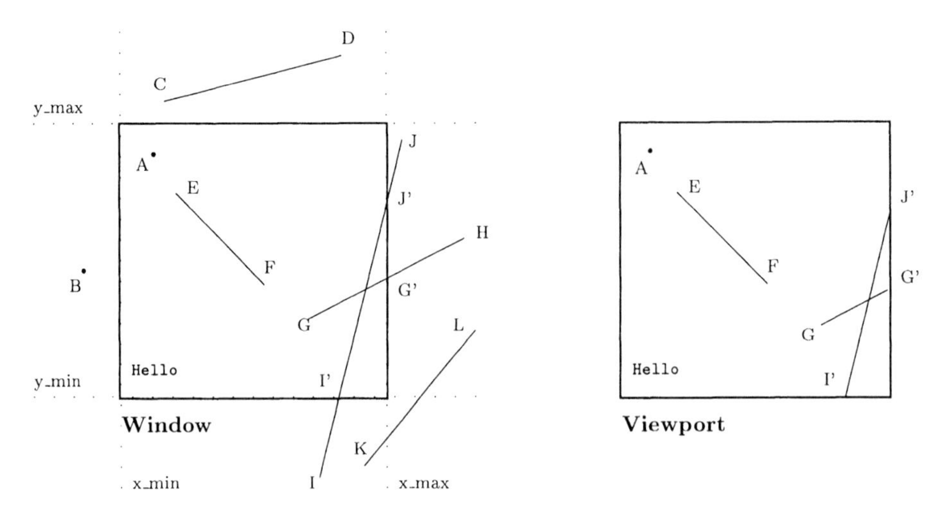

Figure 9.4: Clipping of points and lines

9.2.2.2 The Cohen-Sutherland Algorithm

The primary idea of Cohen and Sutherland (see, e.g., [Foley *et al.*, 1996]) was the infinite stretching of the line segments that delimit the window, thus subdividing the window plane into nine regions (see Table 9.1). Each of these regions is being assigned a so-called *outcode* consisting of four bits:

1001	1000	1010
0001	0000	0010
0101	0100	0110

Table 9.1: The definition of outcodes

- Bit 1 is set to 1, if the region is above the window, else it is set to 0.
- Bit 2 is set to 1, if the region is below the window, else it is set to 0.
- Bit 3 is set to 1, if the region is right of the window, else it is set to 0.
- Bit 4 is set to 1, if the region is left of the window, else it is set to 0.

In order to clip a line, the start point and the end point are first assigned the outcodes of the region in which they are situated (O_S and O_E, respectively).

Four cases can be distinguished:

1. **Both points are situated inside the window:** This is indicated by the *bitwise disjunction* of O_S and O_E yielding $= 0000$. Consequently, this line is accepted and drawn as a whole. In Fig. 9.4, line EF illustrates this case.
2. **Both points are situated above, below, right of, or left of the window:** This is indicated by the *bitwise conjunction* of O_S and O_E yielding $\neq 0000$. Consequently, the line is neglected and not drawn at all, see line CD in Fig. 9.4.
3. **One point is situated within the window, the other outside the window:** This is indicated by the bitwise conjunction of O_S and O_E yielding $= 0000$, the bitwise disjunction yielding $\neq 0000$. This case has to be further investigated; line GH is an example.
4. **Both points are situated outside the window:** This is indicated by the bitwise conjunction of O_S and O_E yielding $= 0000$, the bitwise disjunction yielding $\neq 0000$. This case has to be further investigated in the same way as in case 3; line KL is an example for a line that lies completely outside the window, line IJ is an example for a line that lies partially within the window.

The first two cases cover the major part of the lines in two important cases: If the window is big, many lines are visible as a whole; if the window is small, many lines are completely invisible.

In cases 3 and 4, the following iteration procedure is applied:

The bitwise disjunction of O_S and O_E determines by its 1-bits, which of the window boundaries have to be intersected with the line. Examples: $GH : O_G \vee O_H = 0010 \longrightarrow$ right boundary. $IJ : O_I \vee O_J = 0110 \longrightarrow$ lower and right boundary. If more than one boundary is indicated, one of them can be chosen arbitrarily.

The intersection between the line and the affected boundary line is computed in the following way: One coordinate can be determined from the window boundary without computation, the other is computed by evaluating the equation defining the line, using the first coordinate. Example: $GH : O_G \vee O_H = 0010 \longrightarrow$ intersection with the right boundary $\longrightarrow x' = x_{max} \longrightarrow G' = (x_{max}, y')$.

Thus, the line is split into two segments, one leading from the intersection point into the window, one leading out (*divide-and-conquer approach*). Example: $GH \longrightarrow GG'$, $G'H$.

The line segment leading into the window (GG' in the example) is analyzed further (recursive application of the same procedure), the other segment ($G'H$ in the example) is rejected. Example: $IJ \longrightarrow II'$ (case 2, rejected), $I'J$ (case 3, investigated further); $I'J \longrightarrow I'J'$ (case 1, accepted), $J'J$ (case 2, rejected).

Points that are situated on the boundary are treated consistent (i. e., dimensionally correct) with the rest of the line, see Chapter 8.

Summarizing, the algorithm by Cohen-Sutherland for the clipping of lines can be sketched in the following way:

```
Procedure Clipping (↓ x1, y1, x2, y2, xmin, xmax, ymin, ymax : real)
{ Cohen-Sutherland Clipping Algorithm for line P1 = (x1, y1) to P2 = (x2, y2) }

type outcode = array [1..4] of boolean;
var accept, reject, done: boolean;
var outcode1, outcode2: outcode;    { Outcodes for P1 and P2 }

{ Definitions for Swap, Reject, Check, AcceptCheck and Outcodes go here. }
{ Outcodes takes Clipper's window parameters and a point and returns a 4-bit outcode. }

begin
   accept := false;
   reject := false;
   done := false;

   repeat
      Outcodes (xmin, xmax, ymin, ymax, x1, y1, outcode1);
      Outcodes (xmin, xmax, ymin, ymax, x2, y2, outcode2);
      reject := RejectCheck (outcode1, outcode2);   { check trivial reject }
      if reject then done := true;
      else begin   { possibly accept }
         accept := AcceptCheck (outcode1, outcode2);   { check trivial accept }
         if accept then done := true;
```

```
        else begin    { subdivide line since at most one endpoint is inside }
          { First, if P1 is inside window, exchange points 1 and 2 and their outcodes}
          { to guarantee that P1 is outside window, using Swap }
          if not(outcode1[1] or outcode1[2] or outcode1[3] or outcode1[4]) then
            Swap (x1, x2);
            Swap (y1, y2);
            Swap (outcode1, outcode2);

          { Now perform a subdivision, move P1 to the intersection point; use the }
          { formulas y = y1 + slope · (x - x1) and x = x1 + (1/slope) · (y - y1) }
          if outcode[1] then begin    { divide line at top of window }
            x1 := x1 + (x2 - x1) * (ymax - y1) / (y2 - y1);
            y1 := ymax;
            end
          else if outcode[2] then begin    { divide line at bottom of window }
            x1 := x1 + (x2 - x1) * (ymin - y1) / (y2 - y1);
            y1 := ymin;
            end
          else if outcode[3] then begin    { divide line at right edge of window }
            y1 := y1 + (y2 - y1) * (xmax - x1) / (x2 - x1);
            x1 := xmax;
            end
          else if outcode[4] then begin    { divide line at left edge of window }
            y1 := y1 + (y2 - y1) * (xmin - x1) / (x2 - x1);
            x1 := xmin;
            end
          end    { subdivide }
        end    { possible accept }
    until done;
  if accept then Draw (x1, y1, x2, y2);
  end    { Clipping }
```

The advantage of the Cohen-Sutherland algorithm is an efficient strategy for a complete analysis of total and partial visibility. The disadvantage is that multiplications and divisions are – although reduced in their number – still necessary. This leads to time problems when realizing the algorithm in hardware. The midpoint subdivision algorithm that is presented next overcomes these problems.

9.2.2.3 Midpoint Subdivision

The midpoint subdivision algorithm analyses lines using outcodes in the very same way as the Cohen-Sutherland algorithm until a line is found that can not be either rejected or accepted [Foley *et al.*, 1996]. Such a line is subdivided into two segments of equal length ("midpoint subdivision"), see Fig. 9.5. Both segments are investigated recursively until each of the sub-segments is either rejected or accepted, the idea being similar to binary searching. The smallest segment length to be investigated is defined by the resolution of the viewport hardware.

The computation of the midpoint $M = ((x_1 + x_2)/2, (y_1 + y_2)/2)$ can be performed very efficiently, if integer coordinates are available (e. g., machine coordinates of the graphic hardware): only additions and shift operations are necessary.

The maximum number of iterations is determined by the resolution of the graphic hardware: $\#iterations \leq \lceil \log_2(N) \rceil$, where $N = \max(|x_1 - x_2|, |y_1 - y2|)$. For a resolution of 1024 x 768, the number of iterations consequently is always ≤ 10!

In that way, multiplications and divisions are replaced by a rather low number of additions and shifts, thus improving the performance and enabling efficient hardware implementation.

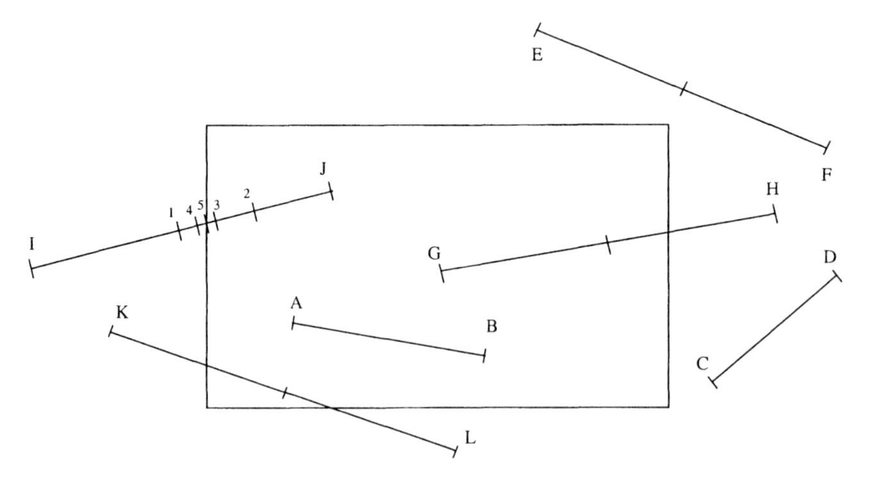

Figure 9.5: Clipping using midpoint subdivision

9.3 Scan Conversion

9.3.1 Scan Conversion of Lines

Modern graphic output devices are – with the exception of pen plotters – raster devices, i. e., they compose pictures out of discrete picture elements (*pixels*). This has the consequence that graphic components like lines and arcs have to be mapped onto the raster grid of the hardware, i. e., the components have to be approximated by a number of such pixels on discrete grid positions. Fig. 9.6 illustrates this situation for lines of different angles. This conversion of objects onto the raster grid is generally called *raster conversion* or *scan conversion*.

The requirements for scan conversion are the following:

- The drawing should be smooth, "staircases" instead of lines should be avoided.
- Delimiting points should be "exact".
- The length or angle of a line should have no influence onto the brightness or color of the line representation.

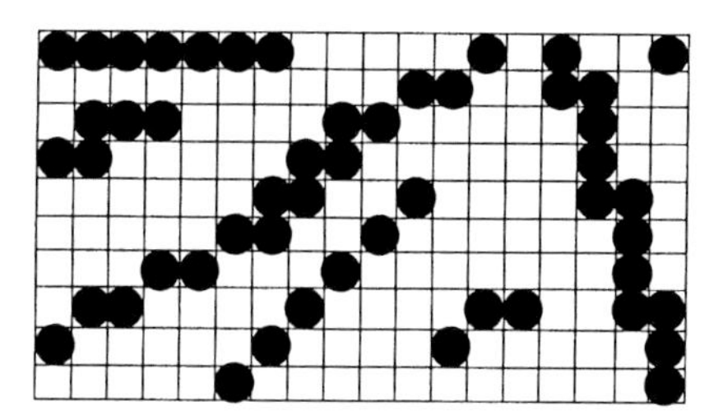

Figure 9.6: Scan conversion of lines

- Performance is crucial since scan conversion is necessary for each line drawn at each time an image is computed.

This – again – leads to conflicting goals: computation time vs. quality! The following algorithms consider both goals trying to describe an acceptable compromise.

9.3.1.1 The Digital Differential Analyzer Algorithm

The rather simple idea of this approach is to round the y-values of the line towards the closest pixel position, see Fig. 9.7.

If a line SE is defined by $S = (x_1, y_1)$ and $E = (x_2, y_2)$,

$$\begin{aligned} dy &= (y_2 - y_1) \\ dx &= (x_2 - x_1) \quad , \text{ assume } x_1 \neq x_2. \end{aligned}$$

and, thus,

$$\begin{aligned} m &= dy/dx, \\ y &= mx + b, \end{aligned}$$

where b is the y-offset of the line from $(0, 0)$.

For the following, let us assume that $-1 \leq m \leq 1$.

Let $dx := 1$; then $m = dy$. Without loss of generality, assume $x_s \leq x_e$. (Let the start point be that with the smaller x-coordinate.) Then,

$$x_{i+1} := x_i + 1,$$

$$y_{i+1} := y_i + m = y_i + dy,$$

thus defining the pixels describing the line in the following way:

$$\text{Pixel}(i) := (x_i, \mathsf{Round}(y_i)).$$

If $m > 1$ or $m < -1$ or m is undefined (i. e., $x_s = x_e$), then x and y change the role in the algorithm.

The following is a sketch of an algorithm realizing the Digital Differential Analyzer idea:

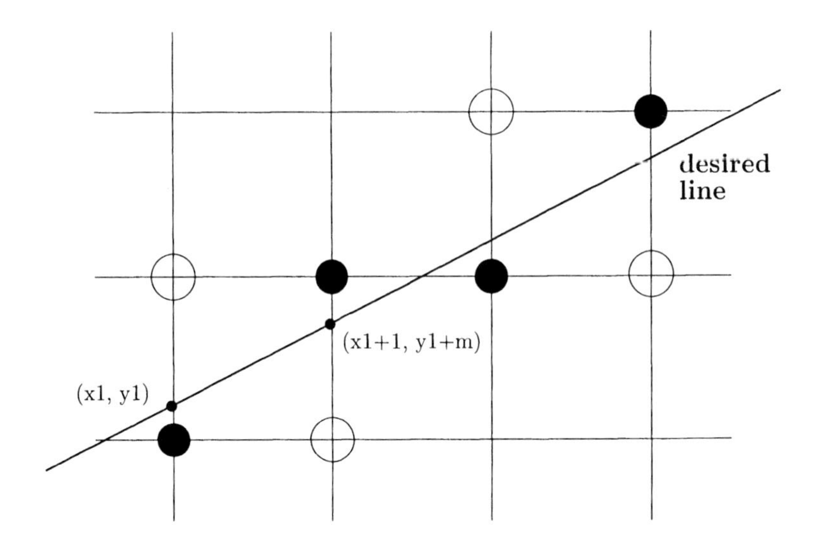

Figure 9.7: The digital differential analyzer algorithm

```
Procedure DdaLine (↓ x1, y1, x2, y2, value : integer)

{ DdaLine assumes slope between +1 and −1 }
{ It draws a line from (x1, y1) to (x2, y2) }
{ value: value to place pixels near line }

var dx, dy, x, y, m: real;

begin
   if x1 <> x2 then begin
      dy := y2 - y1;
      dx := x2 - x1;
      m := dy / dx;
      y := y1;

      for x:=x1 to x2 do begin
         WritePixel (x, Round (y), value);   { sets pixel to value }
         y := y + m;        { step y by slope m }
         end
      end
   { if "line" is really a point, then plot it }
   { otherwise, error }
   else if y1 = y2 then WritePixel (x1, y1, value);
      else Error;

   end    { DdaLine }
```

The disadvantages of this approach are the computations using real numbers (division!) and the Round procedure that is necessary.

9.3.1.2 Bresenham's Algorithm

The following algorithm predates the age of graphics terminals and workstations. It bears the name of J. E. Bresenham, who first described it in 1965 in the context of driving the stepper motors for digital plotters [Foley *et al.*, 1996]. Its efficiency and elegance, however, have maintained the usefulness of this and related algorithms.

The improvement of the scan conversion of lines by Bresenham is based upon the following idea (Fig. 9.8):

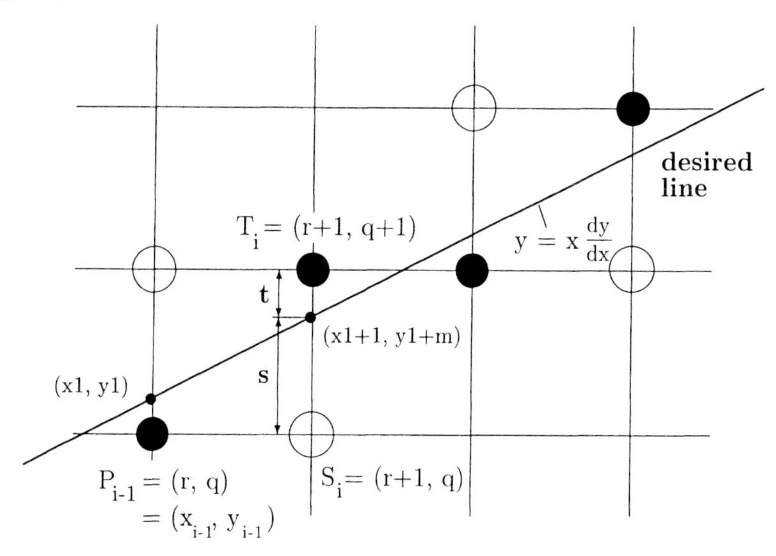

Figure 9.8: Geometry for Bresenham's algorithm

Let us assume that the inclination of the line is in the range $[-1, 1]$.[1]

Assume that pixel $P_{i-1} = (x_{i-1}, y_{i-1})$ is being set. Next, let us set pixel P_i. Let s be the deviation of the line intersection with the ith vertical grid line from the (i-1)th horizontal grid line, let t be the deviation of the line intersection with the ith vertical grid line from the ith horizontal grid line, see Fig. 9.8. Then

$$P_i = S_i = (x_{i-1} + 1, y_{i-1}), \ if \ s - t < 0.$$
$$P_i = T_i = (x_{i-1} + 1, y_{i-1} + 1), \ if \ s - t \geq 0.$$

A control variable d_i is being computed, deciding which case holds. The following procedure concentrates on an efficient computation in order to guarantee a fast decision:

Let us transform the line segment in such a way that the start point (x_0, y_0) is positioned at $(0, 0)$. Then the end point of the line segment is being placed at (dx, dy) with $dx > 0$ and $dy > 0$ and $0 \leq dy/dx \leq 1$. Then, the equation describing the line is $y = (dy/dx)x$.

[1]This means no loss of generality: All other inclinations can be transformed properly by reflecting with respect to coordinate axes and / or the median lines.

In order to simplify the denotation, let us set $r = x_{i-1}$ and $q = y_{i-1}$. Then $P_{i-1} = (r, q)$, $S_i = (r+1, q)$, and $T_i = (r+1, q+1)$.

Then,

$$s + q = (dy/dx)(r+1) = q + 1 - t,$$

and, thus,

$$s = (dy/dx)(r+1) - q$$

and

$$t = q + 1 - (dy/dx)(r+1),$$

Consequently,

$$s - t = 2(dy/dx)(r+1) - 2q - 1,$$

$$dx(s-t) = 2rdy + 2dy - 2qdx - dx,$$

$$dx(s-t) = 2(rdy - qdx) + 2dy - dx.$$

By assumption, $dx > 0$, therefore

$$dx(s-t) \geq 0 \Longleftrightarrow s - t \geq 0$$

and it is sufficient to compute

$$d_i = dx(s-t)$$

in order to decide whether to choose S_i or T_i for P_i. Let us analyze the computation of d_i further:

$$d_i = 2(rdy - qdx) + 2dy - dx.$$

Using $r = x_{i-1}$ and $q = y_{i-1}$, we get

$$d_i = 2x_{i-1}dy - 2y_{i-1}dx + 2dy - dx.$$

Analogously, the computation of d_{i+1} can be derived:

$$d_{i+1} = 2x_i dy - 2y_i dx + 2dy - dx,$$

This results in the difference

$$d_{i+1} - d_i = 2dy(x_i - x_{i-1}) - 2dx(y_i - y_{i-1}).$$

Since $x_i = x_{i-1} + 1$,

$$d_{i+1} = d_i + 2dy - 2dx(y_i - y_{i-1}).$$

This results in a recursive algorithm for the computation of d_i:

1. Initial step: $i = 1$ and $(x_0, y_0) = (0, 0)$. Thus, $d_1 = 2dy - dx$.
2. Case $d_i \geq 0$: $P_i = T_i$, therefore $y_i = y_{i-1} + 1$ and, thus, $d_{i+1} = d_i + 2(dy - dx)$.
3. Case $d_i < 0$: $P_i = S_i$, therefore $y_i = y_{i-1}$ and, thus, $d_{i+1} = d_i + 2dy$.

The following algorithm includes this recursive computation of the control variable d_i:

```
Procedure Bresenham (↓ x1, y1, x2, y2, value : integer)

{ Condition: Abs((y2-y1)/(x2-x1)) ≤ 1; else swap (x1, y1) with (x2, y2) and restart }

var dx, dy, incr1, incr2, d, x, y, xend: integer;

begin
  dx := Abs (x2 - x1);
  dy := Abs (y2 - y1);
  d := 2 * dy - dx;   { initial value for d }
  incr1 := 2 * dy;   { constant used for increment if d < 0 }
  incr2 := 2 * (dy - dx);   { constant used for increment if d ≥ 0}
  if x1 > x2 then begin   { start at point with smaller x }
    x := x2;
    y := y2;
    xend := x1;
    end
  else begin
    x := x1;
    y := y1;
    xend := x2;
    end

  WritePixel (x, y, value);   { first point on line }
  while x < xend do begin
    x := x + 1;
    if d < 0 then   { choose S_i - no change in y}
      d := d + incr1;
    else begin   { choose T_i - y is incremented }
      y := y + 1;
      d := d + incr2;
      end
    WritePixel (x, y, value);   { the selected point near the line }
    end   { while }
  end   { Bresenham }
```

The advantage of Bresenham's algorithm is that it contains only integer additions and integer subtractions and only multiplications by 2 are necessary, which can be realized using the shift operation. In this way, an efficient implementation in hardware is made possible.

9.3.2 Scan Conversion of Circular Arcs

9.3.2.1 Simple Algorithms

The first approach to draw a circular arc is very similar to the Digital Differential Analyzer Algorithm. Let us assume that without loss of generality the center of the circle is the origin,

i. e., the circle is described by the equation $x^2 + y^2 = r^2$ for a given radius r. Thus, for any $x \geq 0$, the corresponding y can be computed by $y = \pm\sqrt{r^2 - x^2}$. The points of the other quadrants can be determined by symmetry.

The (quarter-)circle can be drawn in the following way, see Fig. 9.9:

1. x is being assigned unit values from 0 to r.
2. For each x, its corresponding y-value is computed using the circle equation described above.
3. Pixel $(x, \mathsf{Round}(y))$ is set.

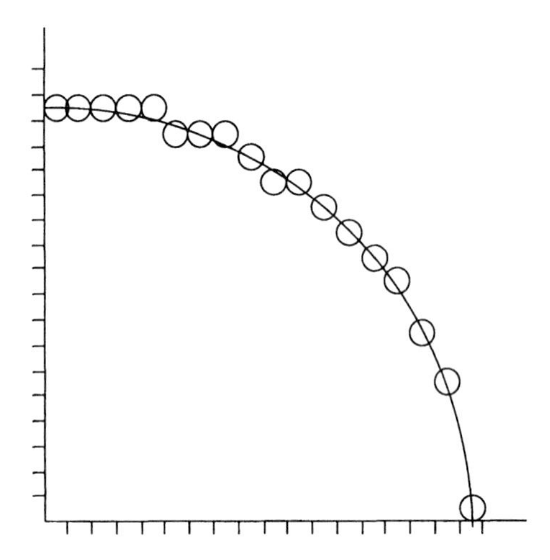

Figure 9.9: A quarter-circle generated with unit steps in x and with y calculated and then rounded

The disadvantages of this approach are the time-consuming operations like multiplication, square root, and rounding, and, on the other hand, a poor graphic quality despite that effort.

The poor graphic quality can be improved by using *polar coordinates*, i. e., an angle α is used as the loop variable, taking values from 0 to 90 degrees, and the corresponding x-values and y-values are computed by $x = r\cos\alpha$ and $y = r\sin\alpha$. The pixel to be set is $(\mathsf{Round}(x), \mathsf{Round}(y))$.

This alternative, however, does not decrease the complexity of computation. Supported by the ideas of Michener, Bresenham applied his concepts for the scan conversion of lines also to circular arcs, yielding a much more efficient algorithm than the simple approaches.

9.3.2.2 The Algorithm of Bresenham-Michener

The first improvement made by this approach is the utilization of the fact that, in order to draw a whole circle, only an eighth of the points have to be computed, the others can be generated by means of symmetry, see Fig. 9.10.

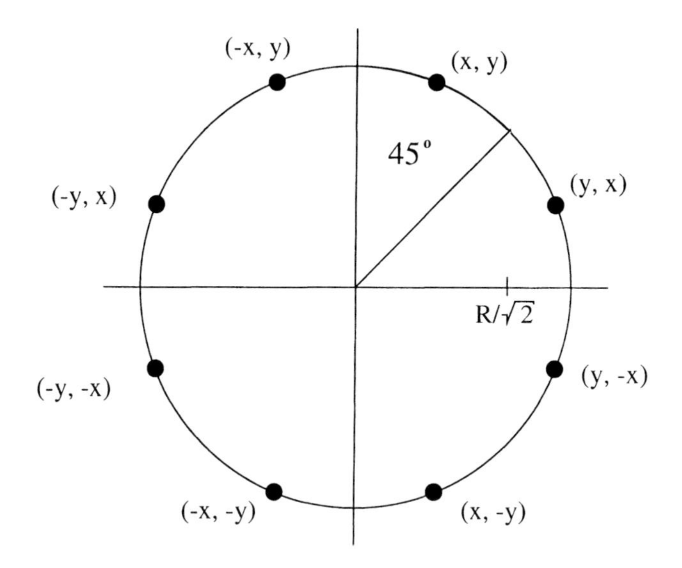

Figure 9.10: Eight symmetrical points on a circle

The basic concept of the algorithm for scan conversion of circular arcs by Bresenham-Michener is very similar to the line case [Foley *et al.*, 1996]: In each step i, that pixel $P_i = (x_i, y_i)$ is set that shows the minimum error $|D(P_i)|$ with $D(P_i) = (x_i^2 + y_i^2) - r^2$.

In Step i, two pixels S_i and T_i have to be considered to be set next, as is depicted in Fig. 9.11. Again, the decision can be made dependent on a control variable d_i.

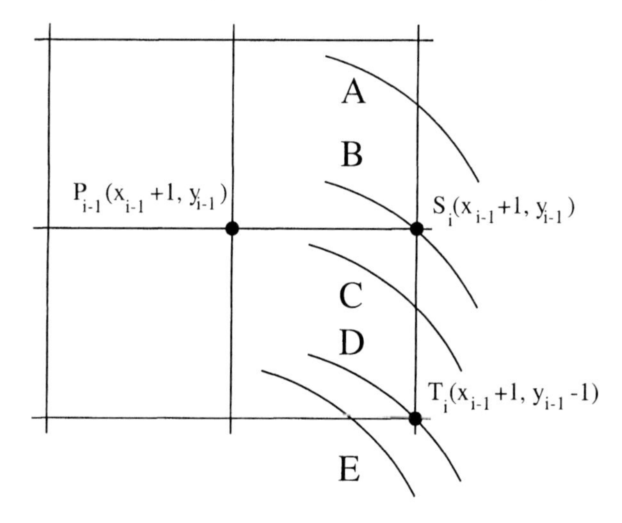

Figure 9.11: Decision points for Bresenham's circle generator

If pixel $P_{i-1} = (x_{i-1}, y_{i-1})$ has been set in Step $i-1$, the two alternatives for pixel P_i are S_i and T_i with

$$D(S_i) = (x_{i-1} + 1)^2 + (y_{i-1})^2 - r^2$$

and

$$D(T_i) = (x_{i-1} + 1)^2 + (y_{i-1} - 1)^2 - r^2.$$

P_i is set to $S_i = (x_{i-1} + 1, y_{i-1})$, if $|D(S_i)| < |D(T_i)|$ and set to $T_i = (x_{i-1} + 1, y_{i-1} - 1)$, if $|D(S_i)| \geq |D(T_i)|$.

$D_i = |D(S_i)| - |D(T_i)|$ would consequently be a suitable control variable, only that its computation is still too complex.

The first improvement is the omission of the absolute function, i.e., $d_i = D(S_i) + D(T_i)$, resulting in S_i to be set, if $d_i < 0$, and T_i to be set, if $d_i \geq 0$. All five alternatives shown in Fig. 9.11 fit into this strategy:

- **Cases A and B**: $D(S_i) \leq 0$, since S_i is within (case A) or on the circle (case B). $D(T_i) < 0$, since T_i is within the circle. Therefore $d_i < 0$ and S_i is set.
- **Case C**: $D(S_i) > 0$, since S_i is not within the circle. $D(T_i) < 0$, since T_i is within the circle. Therefore, if $|D(S_i)| < |D(T_i)|$ then $d_i < 0$ and S_i is set. If $|D(S_i)| \geq |D(T_i)|$ then $d_i \geq 0$ and T_i is set.
- **Cases D and E**: $D(S_i) > 0$ and $D(T_i) \geq 0$. Therefore, $d_i > 0$ and T_i is set.

The remaining multiplications of the computation of d_i can be eliminated by transformations similar to the line algorithm, and one gets the following efficient, recursive algorithm for the computation of the control variable d_i:

1. Initial step: $d_1 = 3 - 2r$.
2. Case $d_i \geq 0$: $P_i = T_i$, therefore $y_i = y_{i-1} - 1$ and, thus, $d_{i+1} = d_i + 4(x_{i-1} - y_{i-1}) + 10$.
3. Case $d_i < 0$: $P_i = S_i$, therefore $y_i = y_{i-1}$ and, thus, $d_{i+1} = d_i + 4x_{i-1} + 6$.

The following algorithm includes this recursive computation of the control variable d_i:

```
Procedure MichCircle (↓ radius, value : integer)

{ assumes center of circle is at origin }

var x, y, d: integer;

begin
   x := 0;
   y := radius;
   d := 3 - 2*radius;
   while x < y do begin
      CirclePoints (x, y, value);
      if d < 0
         then d := d + 4*x + 6;    { select S }
         else begin       { select T – decrement y }
```

```
            d := d + 4*(x - y) + 10;
            y := y - 1;
            end
        x := x + 1;
        end   { while }
    if x = y then CirclePoints (x, y, value);
    end   { MichCircle }
```

Procedure CirclePoints ($\downarrow x, y, value$: **integer**)

```
begin
    WritePixel (x, y, value);
    WritePixel (y, x, value);
    WritePixel (x, -y, value);
    WritePixel (y, -x, value);
    WritePixel (-x, -y, value);
    WritePixel (-y, -x, value);
    WritePixel (-x, y, value);
    WritePixel (-y, x, value);
    end   { CirclePoints }
```

Fig. 9.12 shows a (quarter-)circle computed using the algorithm by Bresenham-Michener.

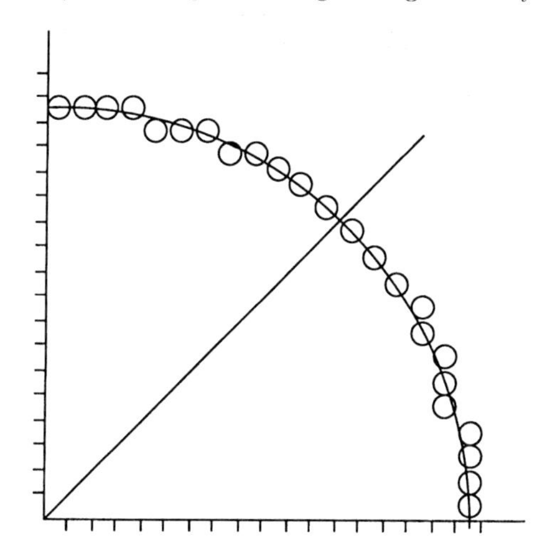

Figure 9.12: Quarter circle generated by the Bresenham-Michener algorithm

9.3.3 Anti-Aliasing

The visualization of graphic objects using raster display has the disadvantage that non-horizontal and non-vertical lines look jagged ("staircase effect"). In order to reduce this problem, one can increase the resolution of the device. This works fine, e. g., for laser printers. However, display resolutions in the range of 1280 x 1024 seem to remain state-of-the-art for

– at least – some more years. In order to improve the visualization quality on devices like displays, too, so-called *anti-aliasing* methods have been developed [Abrash, 1992].

The principle of anti-aliasing is based upon the fact that every line, when drawn in reality (e. g., using a pencil), not only has a specific length, it also has a certain breadth. For anti-aliasing, a breadth is assigned to each line, thus specifying a certain area that is covered by the line, see Fig. 9.13. For each pixel that is covered by this area in part or as a whole, the percentage of coverage is computed. Depending on this percentage, the intensity of coloring this pixel with the line color is varied. Thus, the visualization of the line looks much more smooth and "real".

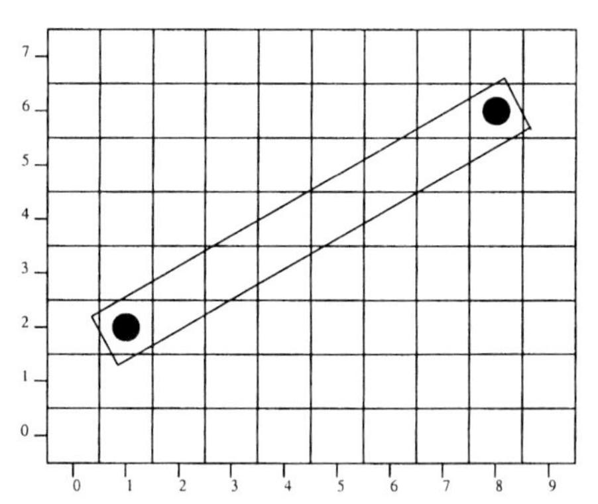

Figure 9.13: Drawing a line of nonzero width

A good illustration example for anti-aliasing is the smooth visualization of characters. As depicted in Fig. 9.14, characters can already be mapped onto the grid using the anti-aliasing technique, thus resulting in a (virtually) better resolution on a color screen than when using just the color - no color duality.

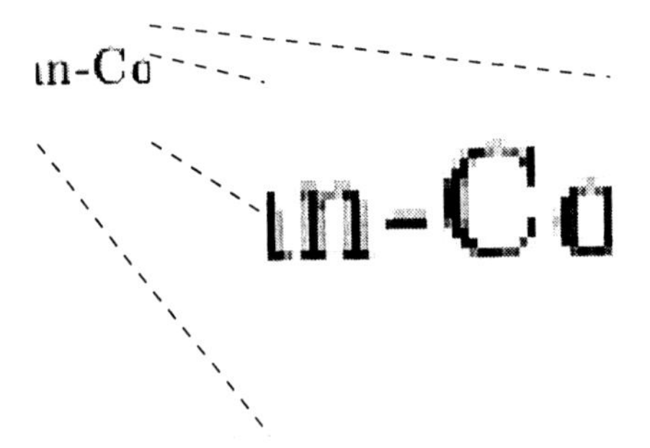

Figure 9.14: Anti-aliasing of characters

Another problem that can be overcome using anti-aliasing is the *unequal intensity problem*, see Fig. 9.15. On a raster display, the brightness of a line is depending on its inclination, because lines with an inclination close to 1 (1st median line) are represented by pixels that are more distant from each other than horizontal or vertical lines. Anti-aliasing solves this problem automatically, because for oblique lines, additional pixels are (partially) set automatically.

The main disadvantages of anti-aliasing can be seen in the example of Fig. 9.14: The sharpness

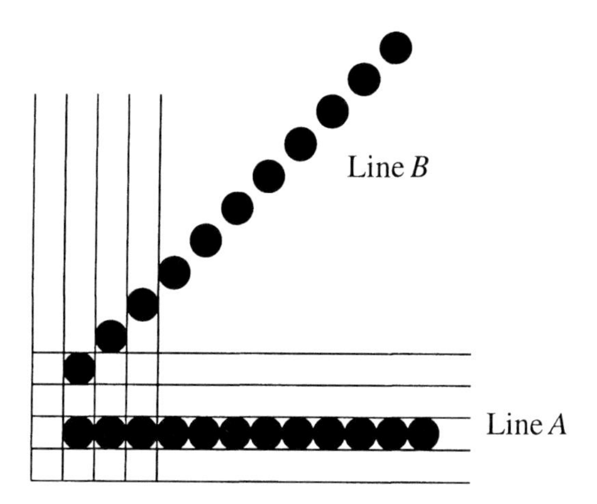

Figure 9.15: The unequal intensity problem

of the image is considerably reduced, so-called *smearing effects* may be quite intense at certain angles, or when several anti-aliased objects are overlaid onto each other.

Furthermore, the anti-aliasing computation is quite time-consuming. A realization in hardware can, therefore, be found quite frequently.

Chapter 10

Generation of Realistic Images

This chapter gives a survey on improving images in order that they look more "realistic" to the user. Techniques to achieve this goal include hidden line / hidden surface removal, rendering methods, and texturing.

10.1 Hidden Line / Hidden Surface Removal

The removal of hidden lines or surfaces is a very common strategy for getting realistic images, besides or additional to parallel projections, perspective constructions, the usage of differently thick lines, and shading.

The strategy of elimination depends on the internal representation of the objects. The data structure of the objects must be a surface model or a volume model, wire frames lack the necessary boundary information. Two different strategies are commonly used:

1. **Surface tests**: Two surfaces are compared with each other and the hidden parts are eliminated. Complexity is proportional to n^2 (n being the number of polygons) and the single computing steps are very time-consuming because of complex geometric relations.
2. **Point tests**: For each pixel it is determined, if a polygon is visible and which one is visible. Complexity is proportional to n * N (N being the number of pixels) with comparatively simple computing operations.

The main problem is the efficiency of these algorithms. Therefore, they are frequently substituted by simple, brute-force algorithms and tests. These, however, may cause wrong decisions in complex situations.

10.1.1 Hidden Line Removal

Objects are represented by the edges of the object surface.

The hidden edges – or parts of the edges – are eliminated, see Fig. 10.1.

For algorithms, we refer to the next section.

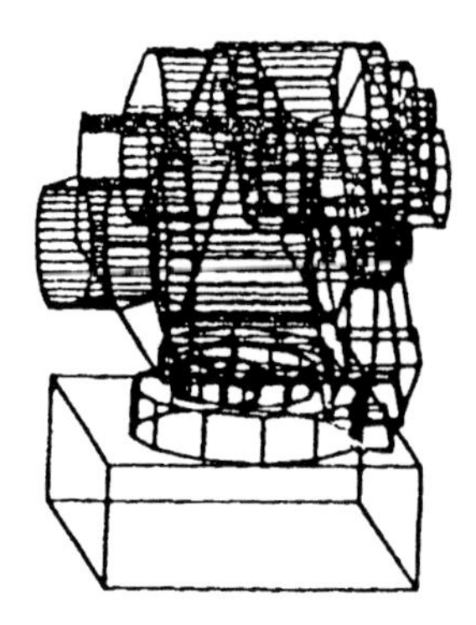 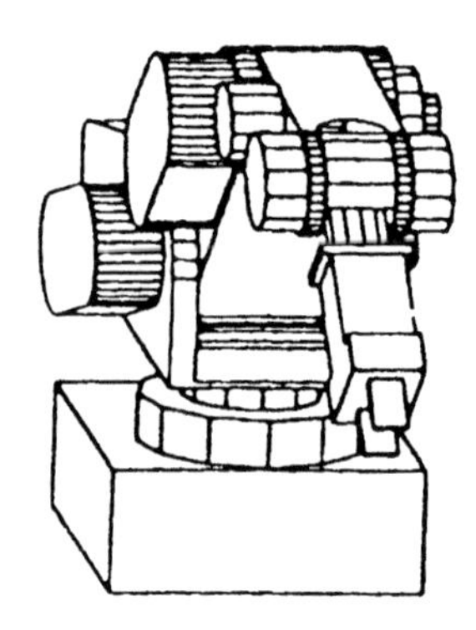

Figure 10.1: The same object without and with hidden lines removed

10.1.2 Hidden Surface Removal

Objects are represented by surfaces and the hidden surfaces or surface parts are eliminated.

10.1.2.1 Facet Intersection

By its structure, this algorithm is a point test (area subdivision algorithm), although not every single point is examined.

Similar to the quadtree / octree concept, the representation surface is divided into rectangular windows. Each window is divided further, until the visibility decision can be made straightforwardly. If this process generates windows with the size of a pixel, it also stops.

For the decision process, the knowledge is utilized that only four different relations between a polygon and a window are possible, see Fig. 10.2.

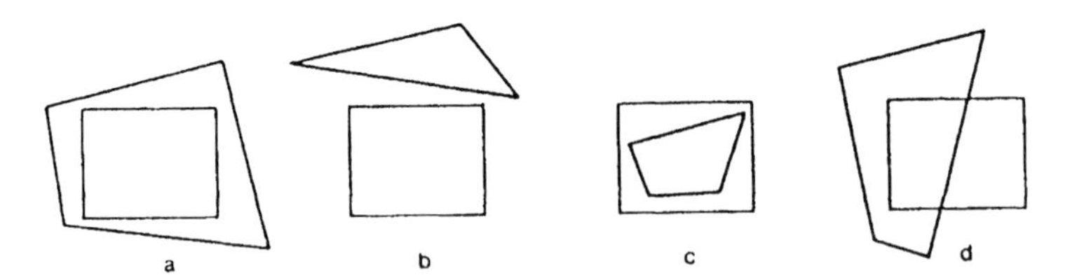

Figure 10.2: Possible relations between window and polygon

The following cases can be decided immediately:

1. The window is surrounded by only one polygon and disjoint with all others (a). It is filled with the color of the polygon.
2. The window is disjoint with all polygons (b). It is filled with the color of the background.
3. The window intersects with only one polygon (d), or surrounds it (c), and is disjoint

with all others. The surrounded part of the polygon is painted with the respective color and the rest with the color of the background.

4. There is a polygon that surrounds all other polygons in this examined window and also surrounds this window (a). The whole background can be filled with the color of this polygon, the surrounded polygons are drawn appropriately.

5. If the window size is smaller or equal to the display resolution, the pixel is painted in the color of the front polygon.

In all other cases the window is divided, until one of the above cases holds.

10.1.2.2 The Painter's Algorithm

The idea of this algorithm is to sort all polygons from backward to forward by their z-coordinates (depth sort algorithm). Then all polygons are drawn in this sequence, so that the polygon with the smallest distance from the view point is the last one to be drawn ("painted" over the others). [Foley *et al.*, 1996]

The main three steps of the algorithm are:

1. Sort all polygons relating to their z-value. Transformations have to assure that the view plane is parallel to the XY-plane, and that the z-axis extends positively into the viewing direction.

2. Solve all ambiguities that come up when the z-extents of the polygons overlap, see Fig. 10.3 for some tricky cases.

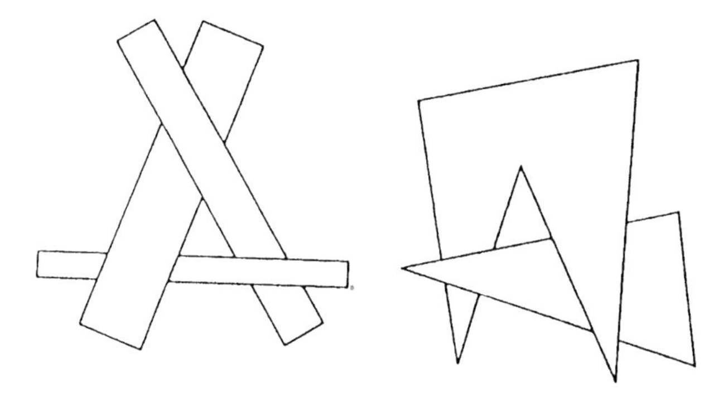

Figure 10.3: Cases of ambiguity when sorting polygons with respect to depth

- The following five tests of growing complexity are performed to check this ambiguity. If one test succeeds, the polygons are already correctly sorted:
  (a) Check whether the x-extents of two polygons do not overlap.
  (b) Check analogously for y-extents. (Steps 1 and 2 are sometimes called "Minimax Test".)

(c) If all points of the back polygon are behind the plane defined by the polygon in front, they are sorted correctly.
(d) If all points of the front polygon are in front of the plane defined by the back polygon, they are sorted correctly.
(e) If the images of both polygons do not overlap, the sequence does not matter, therefore they are sorted correctly.

- If all tests fail, then the sequence is wrong. The two polygons are exchanged and the tests start again.
- In order to prevent never ending exchanges, a marked polygon must not be moved once more. In this case, one polygon is intersected with the plane defined by the other polygon. The new polygons gained this way replace the old polygon in the polygon list. Then the tests start again.

3. Enter the polygons into the viewport from backward to forward. Thus, all hidden parts are being "painted over".

10.1.2.3 Z-Buffering

For *z-buffering* [Ward, 1997], in addition to the viewport one needs a buffer for z-values for each pixel. First of all the viewport is initialized with the background color and the z-buffer with the highest value that can be represented. Without sorting every polygon is entered into the viewport in two steps:

1. Compute the polygon depth z(x,y) of the point (x,y).
2. If z(x,y) is smaller than the z-buffer value at this point (x,y), then
   - Enter z(x,y) into the z-buffer.
   - Enter the polygon color into the viewport.

The advantages of this algorithm are the easy implementation and the high efficiency in case of many polygons. However, one needs a large amount of memory designated only for this purpose. (For example, 4 MB are necessary for a 1024 x 1024 image and a 4-byte word for each z-buffer value.)

10.1.2.4 Alpha-Buffering

A modification of the z-buffer algorithm is the *alpha-buffer* method [Ward, 1997]:

- The whole scene is subdivided into regions that are monotonously sorted with respect to depth.
- The single regions are treated one after the other starting with the one upfront.
- Inside each region one estimates the correct visibility with a fitting method.
- First, the region in front is entered into the viewport. For each pixel that has been assigned a color the corresponding alpha-buffer (1 bit) is marked.

- Subsequently, the further regions are entered into the viewport. Only pixels with an unmarked alpha-buffer need to be treated for each region.

This algorithm shows even better performance than the z-buffer algorithm, particularly when appropriate regions are selected.

10.2 Rendering

The rendering process is the last step of image generation, in which the computer transfers the 3D defined model of an object or a whole scene into 2D image representation. Other important components of rendering, besides the algorithms themselves, are:

- **Motion blur:** This technique imitates blur images of objects in motion like one is used to from movies.
- **Composition:** The goal of this technique is to divide a complex image into image parts, like background, building, and object in front of building. In this example, for many applications, it may not be necessary to compute the whole image but only the object in front of the building.
- **Time:** The computing of a single image may take some hours, even on a modern supercomputer. Important determinants are:
  - addressable resolution,
  - number of objects of a scene,
  - number of polygons an object representation consists of,
  - program type for image generation,
  - class of computer system (PC, workstation, supercomputer).

In the following sections, we will sketch the most important contemporary rendering techniques. For more in-depth details, we refer to [Foley *et al.*, 1996].

10.2.1 Shading

The techniques discussed till now allow the representation of objects as colored hidden-line graphics, optionally with colors assigned to the surface polygons. In order to get a realistic image, however, further characteristics of a scenario have to be considered, like the direction and properties of the light source, the surface characteristics of the objects, or the luminosity and transparency of the objects. These influences result in the fact that in reality almost no facet is really of the same color all over the facet. Shadings, shades, reflections and transparent images have an influence on the visual impression of an object. In order to model shaded objects, we first need to investigate the basics of visualization and lighting theory.

10.2.1.1 Shading Functions

The goal of this chapter is to compute the main factors that influence the degree of shading of an object at any point of its surface. Since the assignment of a shading value generally is a rather approximate process due to the restricted number of colors available on a computer system (even 16M is rather restricted in color theory!), one goal of a usable shading function is the simplicity of the function. Only simple functions enable efficient computations. Of course, the shading function should always lead to an acceptable approximation of the "real" shading.

- **Question 1: How to get the shading angle?**

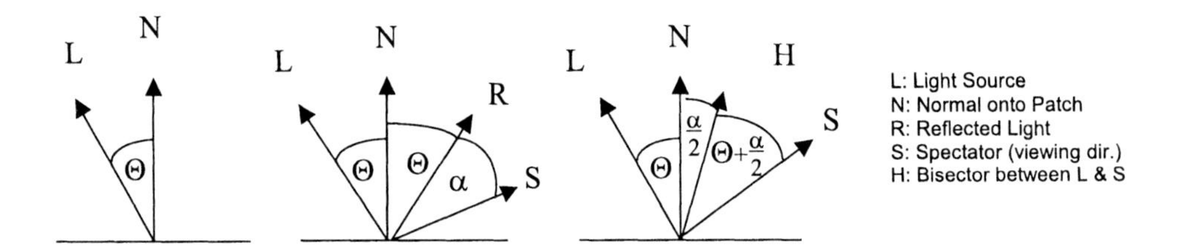

Figure 10.4: Computing the shading angle

The angle of the light source θ (Fig. 10.4) is determined by

$$\cos\theta = \frac{N \cdot L}{|N| * |L|}.$$

The difference of the reflection angle (determined by R) and the viewing angle (determined by S) determines the shading angle α (Fig. 10.4):

$$\cos\alpha = \frac{R \cdot S}{|R| * |S|}.$$

Alternatively, α can be determined in the following way:

$$H = \frac{L + S}{2},$$

$$\cos\frac{\alpha}{2} = \frac{H \cdot N}{|H| * |N|}.$$

- **Question 2: How to get the normal of a surface patch?**

The simplest approach is to compute the weighed average N' of all normal vectors $N_i = (x_i, y_i, z_i)$ (If $i = n$ then $j = 1$ else $j = i + 1$):

$$N'_x = \sum_{i=1}^{n} \frac{(y_i - y_j)(z_i - z_j)}{|N_i| * |N_j|}$$

$$N'_y = \sum_{i=1}^{n} \frac{(z_i - z_j)(x_i - x_j)}{|N_i| * |N_j|}$$

$$N'_z = \sum_{i=1}^{n} \frac{(x_i - x_j)(y_i - y_j)}{|N_i| * |N_j|}$$

For more sophisticated normal computations, we refer to the discussion of shading types later in this section.

- **Question 3: How to get the color and intensity?**

  The computation of the color intensity is defined by Lambert's Law:

$$I = I_l * k_d * \cos\theta$$

where

I	...	reflected intensity
I_l	...	brightness of source of light
k_d	...	reflection constant
θ	...	angle between direction of light ray and normal on polygon

This model creates objects with a dull surface. Parts of the surface being in the shadow of the object are black. This is not very realistic, because in the real world ambient light (light of no fixed direction) brightens the whole object. Thus, we add ambient light in the following way:

$$I = I_a * k_a + I_l * k_d * \cos\theta$$

where

I_a	...	ambient intensity
k_a	...	ambient reflection constant ($0 \le k_a \le 1$)

Still, planes of the same orientation, but with different distance, have the same brightness. Therefore, we weigh the non-ambient light with its distance d:

$$I = I_a * k_a + \frac{I_l * k_d * \cos\theta}{d^2}$$

With this formula the intensity changes unrealistically fast in case of near objects. A better approximation to reality is:

$$I = I_a * k_a + \frac{I_l * k_d * \cos\theta}{d + k}$$

- **Question 4: How to add reflection characteristics?**

  Till now, we only considered rough, undirected surfaces. A mirror, for example, is in complete contrast to this modeling scheme. In a mirror, light is only visible intensively, if the user's eye is more or less exactly in the direction of the reflected ray. Therefore, we add a reflection curve to our formula:

$$I_s = I_l * \omega(i, \lambda) * \cos^n \alpha$$

where

$\omega(i, \lambda)$	...	reflection curve
i	...	angle of incidence
λ	...	wave-length

Reflection curves are quite complex. In most cases, they are therefore substituted by a suitable constant value k_s:

$$I = I_a k_a + \frac{i_l}{d+k} * (k_d * \cos\theta + k_s * \cos^n \alpha)$$

and, summarizing, we get the function for shading intensity,

$$I = I_a k_a + \sum_{j=1}^{n} \frac{i_{l,j}}{d_j + k} * (k_d * \cos\theta_j + k_s * \cos^n \alpha_j).$$

10.2.1.2 Shading Types

Constant Shading

Most image representations of geometric models are based on polygons, whose surface consists of planar facets. Using the shading function defined above, each of the facets can be assigned a single color, since the normal is considered to be constant on the whole facet, see Fig. 10.5. Thus, each polygon is painted in a uniform color. The brightness depends on the lighting direction. This shading type is also-called Polygonal Shading or Lambert Shading.

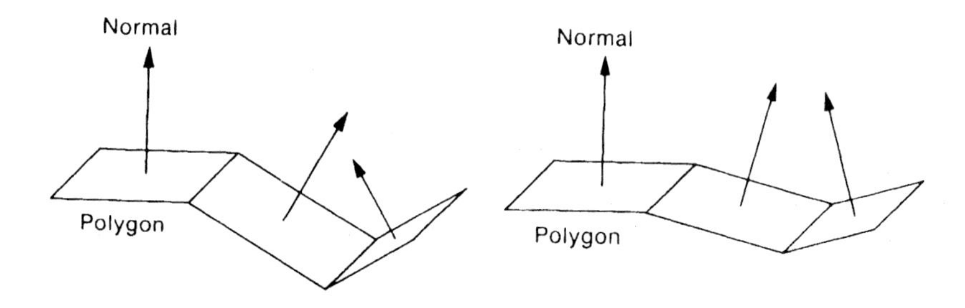

Figure 10.5: The normal is considered to be constant for the whole facet in constant shading

The disadvantage of Constant Shading is that, whereas the original object may be smooth, artificial edges are being introduced at the facets' boundary. Thus, smooth surfaces look rather edgy using this shading method. An example for constant shading is given in Fig. 10.6.

Gouraud Shading

Gouraud had the idea to compute a normal on each corner of a facet, defined by the average of all adjoining facets of the corner. The color values are then computed and interpolated along the polygon edges. This causes uniform and soft color changes. The facet image disappears and this algorithm already causes the impression of a round, smooth surface of, e. g., a sphere. An example for Gouraud shading is given in Fig. 10.7.

Figure 10.6: Image rendered using constant shading

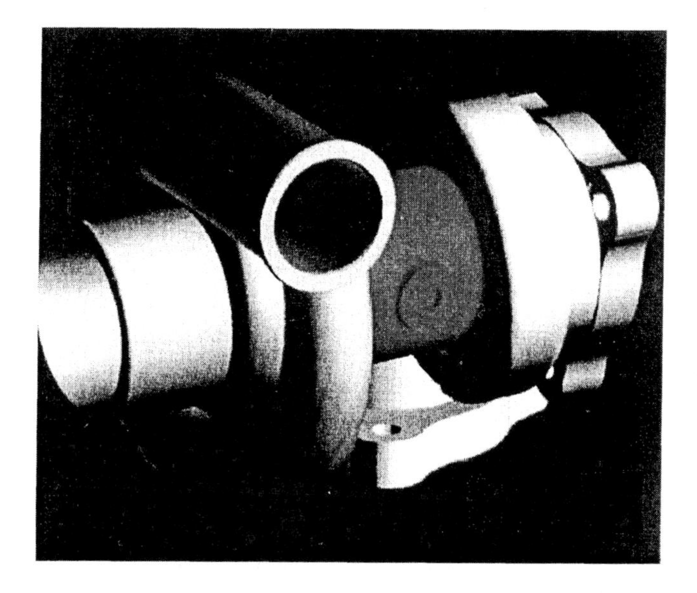

Figure 10.7: Image rendered using Gouraud shading

Phong Shading

Using Gouraud's method, material properties cannot be included into the shading output except at the corners of the facets. Highlighting spots of a glossy metal ball, for example, cannot be visualized. Phong improved the Gouraud shading method by including the three material properties ambience, diffusion and reflection. This algorithm does not interpolate the color intensities; it interpolates the normals along the polygon edges and computes a shading value from these normals for each pixel individually. The highlighting additionally accentuates the three-dimensionality of the image. An example for Phong shading is given in Fig. 10.8.

Phong Shading results in the most realistic pictures of all three methods. The computation time, however, also increases in comparison with Gouraud Shading and – the fastest, showing lowest quality – Constant Shading. Therefore, quite frequently a mixture between Gouraud Shading and Phong Shading is being used.

10.2.2 Ray Tracing

The ray tracing algorithm computes for each pixel its color and its intensity. The principle is to pursue each single light ray and compute the changes of the ray, if it hits any objects. The property of the first point that is hit by the ray determines how the pixel is colored. In order to model reflection, at each point that is hit by the ray, additional "reflection rays" are generated and pursued. This algorithm requires at least a three-dimensional plane model. It considers reflection, refraction and shadowing (reflecting and transparent objects).

Three kinds of input take influence on the result:

- the properties of the objects in the scene, their surface (smooth or rough), their color, if they are transparent or not and their volumetric weight;

- the description of the light source and a possible general brightness, described by the position, direction and intensity of the light.
- time, if moving objects have to be represented.

Problems of this algorithm are the time needed to compute the intersections and that the intersection formula can not always be stated explicitly (e. g., for Bézier Surfaces). An example for ray tracing is given in Fig. 10.9.

Figure 10.9: Example for a ray traced image

10.2.3 Radiosity

This technique, additional to ray tracing, considers diffuse lighting between the objects, too. The process that simulates the interaction of light rays of diffuse reflecting surfaces is called *radiosity*. This model is closest to reality, because it also considers the maintenance of the energy transported by the light.

The main assumption is that all radiation and reflection processes are ideal diffuse reflectors in the meaning of Lambert's law. This means that after a reflection on a surface the direction of the ray vanishes. The radiosity is the sum of the radiated energy (on it's own) and reflected energy and is equivalent to the whole emitted energy.

With this algorithm the modeling of planar light sources is possible, too. The main disadvantage of the radiosity method is that sharp edges tend to get blurred by the radiosity patches and the sharpness of the image is considerably reduced. An example for radiosity is given in Fig. 10.10.

The most promising approach currently seems to be the combination of radiosity and ray tracing algorithms. The realization of an efficient combination, however, is still a research topic.

Figure 10.10: Example for a radiosity-generated image

10.3 Texturing

If objects with irregular and very detailed surfaces (orange, golf ball, human skin, etc.; see Fig. 10.11) are modeled and the image only has a smooth surface, it gives an artificial impression to the human eye. If the modeled object is given a detailed structure, the human eye gets much more information, and this helps to get a more realistic impression.

Figure 10.11: Texturing of irregular surfaces (orange)

There exist four different techniques to assign a surface structure to a 3D-object:

1. texture mapping,
2. bump mapping,
3. solid texturing, and

4. reflection mapping.

10.3.1 Texture Mapping

This technique transforms two-dimensional pictures or patterns onto three-dimensional objects as surface structures (called *textures*). The distortion of the texture, caused by curvatures of the object, is considered by the algorithm. This transformation is called *mapping*.

Consistent textures around a solid object (e. g., wood), however, cause some problems, because the consistency of the solid object can not be integrated into the rules for the surface description.

10.3.2 Bump Mapping

For certain objects it is possible to generate three-dimensional surface structures. A classical example is the structure of a golf ball. This texture-method represents the effect of irregular surfaces during the computation of the brightness of each point.

By changing the local behavior of reflection, a virtual roughness of the surface is simulated. This happens for each point following statistic rules. In this way, differently structured bright and dark points are created on the surface, still "mainly" following the given surface structure.

10.3.3 Solid Texturing

This technique computes the texture directly "into" the object. The computed textures become part of the lighting model. Thus, the object itself contains the texture. Three-dimensional textures are generated in such a way that every point of the space frame is given reflection and transparency properties by mathematical algorithms.

The advantage of solid texturing is its flexibility. Shape and texture are independent, and the data basis is much smaller in comparison to texture mapping. Texture rules imposed by the solid object (e. g., the pattern of wood) are automatically fulfilled by the algorithm. The main disadvantage of solid texturing is the high complexity of the property algorithms.

10.3.4 Reflection Mapping

Reflection mapping, in its optical appearance, is a simplified ray tracing method (sometimes called "poor man's ray tracing"). The surface of an object is considered to have mirroring properties. A single reflection of the rays of the environment of the object maps this environment onto the object. If the object moves, the mapped environment changes accordingly, determined by the position of the object and the reflection angle. The properties of the object itself are considered by the algorithm only in a very restricted way.

Part IV

Simulation

Chapter 11

Manufacturing Simulation

This chapter introduces the concept of simulating a system, focusing on graphic simulation. Different models for and aims of simulation are discussed. Based upon simulation techniques, the verification of tasks and products is sketched together with the possibility of generating robot / NC code automatically.

11.1 Simulation Techniques

A *simulation* is a model of a system [Miller, 1985]. More exactly, simulation is a reproduction of a dynamic (temporal variable) process (model) of a real world phenomenon, in order to arrive at conclusions that are transferable to reality. This scientific process of problem solution is a cyclic process (see Fig. 11.1). The process starts with a real system. This can be an existent, planned, or artificial system. In the modeling state one transfers the real world into a model world. In order to gain a modeled world one makes some restrictions, abstractions, and idealizations, e. g. to simulate air pollution one does not take infinitesimally small boxes to represent the atmosphere of a town. The model is described in a formal (mathematical) language. Then one is able to solve the problem in two ways,

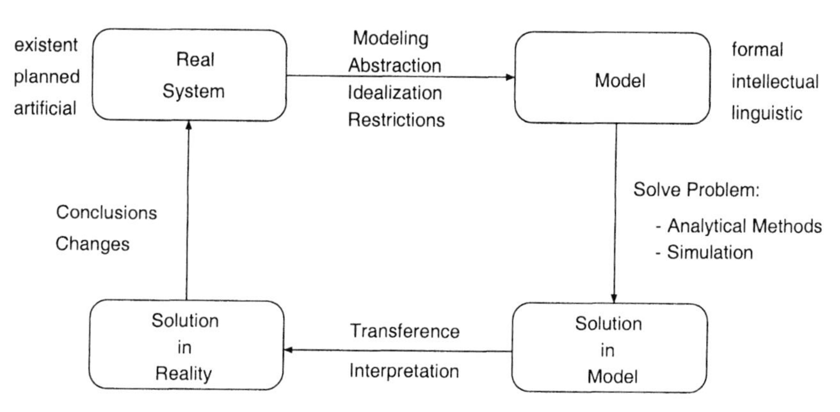

Figure 11.1: A simulation cycle

1. analytical methods (closed form) and
2. simulation (reconstruction of real behavior).

One interprets the solution(s) of the model and transfer(s) them back into real life. At this state a process of reengineering starts. The new conclusions bring a new point of view to the problem and possibly change the real system. This closes the simulation cycle.

11.1.1 Types of Simulation Models

Essentially simulation models can be categorized into

- physical simulation models and
- computer simulation models.

Physical simulation models are actual, physical re-implementations of real world systems, that can either be built more easily or be handled and experimented with more easily. A famous example of a physical simulation model is the wind tunnel. In computer simulation analog or digital computer programs are employed as simulation models.

One can classify simulation models depending on time or changes (see Fig. 11.2). This means, when an amount of time is passed or a significant change appears, the model updates its state. These updates can be continuous or discrete. For a factory automation simulation system one uses discrete changes in combination with continuous time, the so-called *discrete event simulation* (*DEVS*) [Zeigler, 1984].

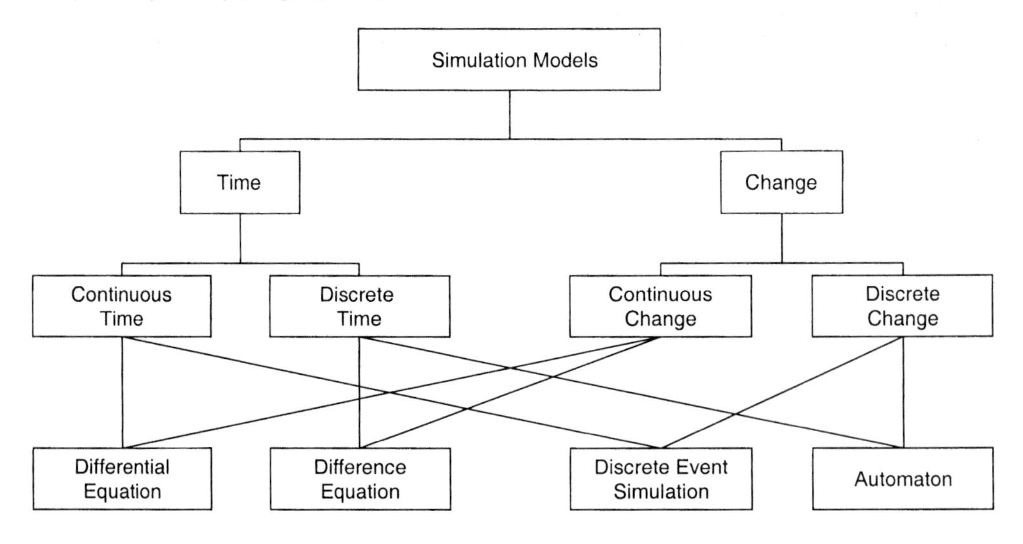

Figure 11.2: Types of simulation models

11.1.2 Applications of Simulation Models

The general nature of simulation offers a wide range of possible applications for different decision making tasks. We define four levels in which simulation models can be employed (see [Pichler and Schwärzel, 1992]):

- explanatory,
- forecast,
- improvement, and
- design.

In the first level, simulation is used to get more insight into the operation and behavior of the real world, especially in business applications. In the forecast level one investigates the behavior of an existing system through a simulation model in the future (e. g. weather forecast). In the third level one uses a model of an existing system to improve it. Several scenarios are built to gain information about the whys and wherefores. The best realizable alternative will be implemented (e. g. assembly lines). No real system exists in the design level. The models serve as design blueprints that are tested by simulation (e. g. traffic simulation, chip production, prototyping, etc.).

Computer simulation has acquired its position as an important tool in many scientific disciplines. The field of applications reaches from continuous to discrete simulation.

11.2 Simulation in Factory Automation

Factory automation simulation refers to the process of developing a model of a work cell, encoding the model into a simulation language, and then utilizing the computerized model to analyze the behavior of the work cell over time [Vig *et al.*, 1988].

Such system models represent the important details on the performance of the system in a way that permits them to be easily manipulated. However, depending on the goal of the simulation and on the data that is available, a broad variety of "simulation" techniques have been developed. The most important types are the following:

- **Network queuing simulation**: Network queuing theory and mean value analysis provide good rough-cut analysis for cell design questions [Winkowski, 1989]. However, they do not include a detailed time history needed to examine decisions made on a day-to-day basis.
- **Discrete event simulation, FMS simulation**: This type of computer simulation estimates the system characteristics of a cell model for specific operation conditions by accumulating statistical data on it over a period of time [Law and Kelton, 1982].
- **Material handling simulation, logistics simulation**: Such a simulation system enables the user to build a logistics system on the computer that serves as a pilot plant (see, e. g., [Waterbury, 1983]). The model can then be tested by running the

simulation. In this way, answers can be provided to questions such as: What will happen if production volumes double next year? Will the plant still operate if the main conveyor goes down?

- **Dynamic process simulation**: Today's increasingly complex industrial process systems are highly sensitive to changes, with minor alterations often creating large differences in the performance of the production. Modeling these processes using computer based simulation (typically by characterizing the system by a set of differential equations; see, e. g., [Fang *et al.*, 1991]) allows the evaluation of alternatives without disrupting the existing process or installing new equipment.
- **Graphic simulation**: The geometric and kinematic description of the components of a manufacturing cell is used to build up a scenario that can be visualized on a computer screen graphically. The modeled task is then used to animate this scenario. Automatic analysis and verification functions may accompany the simulation.

Since our goal is to analyze the geometric and kinematic behavior of mechanisms within a work cell, our simulation is based upon geometric and graphic data. Thus, in the following we will concentrate on *graphic simulation*. If not stated differently, we also use the single term simulation as a substitute for graphic simulation.

11.3 Machining Simulation, Manufacturing Simulation, and Production Simulation

Throughout this chapter, our focus is on so-called *manufacturing simulation systems*, particularly on using them for determining collisions within a work cell caused by an erroneous task. Manufacturing simulation systems allow the efficient modeling, simulation and programming of work cells with robots, NC machines, and (small) production lines. They can be – on the one hand – delimited from *machining simulation systems*, where the simulation of the machining process on a workpiece and the actions "within" an NC machine are in the focus of research. On the other hand, manufacturing simulation systems can be delimited from *production simulation systems* and *production planning systems*, where problems of logistics and the communication within a production line or a production plant are in the foreground.

The major part of the manufacturing simulation systems has been designed in order to simulate (and program) industrial robots because of their rather high task complexity. Consequently, the term *robot simulation systems* can be found very frequently in the literature (see [Rennau and Schnitzler, 1984] for an early survey). An NC machine can be considered as some kind of robot with a restricted freedom of motion, because it generally has less degrees of freedom than an ordinary robot and its joints are restricted to very special types of motions (e. g., Cartesian motions with the same velocity in all three dimensions). Therefore, we consider the terms *robot simulation* and *manufacturing simulation* more or less synonymously.

11.4 History of Graphic Simulation Systems

The concept of graphic manufacturing simulation is not new: Its roots are the origins of computer aided design and modeling in the early 1960s. Subsequently, it has developed from the robot-time-and-motion techniques of the late 1970s to the current animation systems with automatic offline programming capabilities for multi-mechanism work cells. In the following, we list the major milestones of this development:

1. At Kearney & Trecker, Inc., USA, a computer simulation for flexible manufacturing was developed in 1971. The goal of this system was to study all components within a flexible manufacturing system and their interactions with control algorithms. Graphic information was used, e. g., for depicting load curves of specific cells and work units, or for highlighting system bottlenecks.
2. In 1978, General Motors and Boeing confirmed the usefulness of CAD / CAM technology and described how to bridge the gap between CAD and CAM.
3. The first attempt to model robot behavior was the application of time-and-motion methodologies to give predictions of cycle times. The Robot Time and Motion (RTM) approach was first presented in 1978 by R. Paul and S. Nof of Purdue University, USA. The method based upon segmenting the robot motions and adding up times associated with all operations of a task.
4. In the beginning of the 1980s, a simulation system called GRASP (General Robot Arm Simulation Program) was developed by S. Derby at Rensselaer Polytechnic Institute, USA. GRASP can be used to simulate work cells and, additionally, analyze the resulting motions of the single joints of the mechanisms involved.
5. The system PLACE (Positioner Layout And Cell Evaluator) by McDonnell Douglas was the first to include a library of closed form solutions for the inverse kinematics of mechanisms. Previously, the inverse kinematics had to be approximated by numerical iteration methods (e. g., in ROBOTSIM by CALMA, Inc.).

Today, more than 100 systems for graphic manufacturing simulation are available, either as research systems or already as commercial products.

11.5 Characteristics of Graphic Simulation Systems

Historically, manufacturing plants have been satisfactorily designed and built without the benefit of computer simulation. If simulation is to be recognized as a viable management tool, it must either provide better manufacturing designs or offer cost benefits in engineering designs or installation efforts.

As stated in [Rembold and Dillmann, 1986], the increasing use of offline programming systems for the application of industrial robots requires that checking and testing of the generated control information must be possible. The graphic simulation of the robot motion is a tool to generate fail-safe control programs by offline programming. With graphic simulation the motion of the robot is displayed and monitored on a computer screen. In addition the necessary stations of the manufacturing cell are represented. The graphic system can not only be used to check the effects of the program, but can also generate control data for the design and selection of robots as well as for the planning of a manufacturing cell.

In the design of work cells, up to 80 % of the time required for implementation is spent on cell design and programming [Mills, 1985]. This time can be substantially reduced by the use of simulation, consequently reducing production costs considerably [Wagner, 1989]. Several factors make this technique one of today's "hottest" industrial technologies. The most important of these factors are:

1. The development of powerful new microcomputer workstations and simulation software makes simulation affordable for almost any engineering project.
2. The advances in solid modeling and computer graphics provide animation. So designers can visualize the designed systems in operation.
3. The general industry effort to link "islands of automation" by means of computer networks leads to flexible manufacturing systems whose complexity require simulation analysis.
4. Techniques are available to transfer simulation results into offline programming for industrial robots.

Based on the evaluation of a number of installations of simulation systems, the following main benefits of simulation have been identified [Pritsker and Alan, 1984]:

- **Versatility**: Computer modeling may be used to represent a wide variety of real-world systems.
- **Flexibility**: Computer models may be easily altered to represent different situations or updated information.
- **Cost effectiveness**: Experiments using computer simulation enable the performance of a system to be reliably investigated without building the physical system.
- **Non-disruption**: Simulation experiments permit a system to be designed, redesigned, and analyzed without disrupting any existing system.
- **Exhaustion**: Simulation experiments may be performed under every conceivable set of system conditions, parameters, or operating characteristics.

11.6 Requirements for a Kinematic Simulation System

In order to obtain a wide range of functionalities, a kinematic simulation system has to fulfill the following basic requirements [Straßer and Wahl, 1995]:

- **Geometric modeling**: The robot arm and each component of the robot work cell have to be modeled. This modeling consists of various steps depending on the type of objects that have to be modeled, or the purpose of simulation.

- **Kinematic modeling**: A kinematic description is necessary for all mechanical parts that are able to perform any kind of motion; typical examples are the complete robot arm or any other mechanical device such as a part feeder.
- **Collision detection and avoidance**: A sensor based model helps to detect collision and is a very necessary part to improve safety and reduce damage in a work cell.
- **Graphic animation**: In times of virtual reality graphic animation plays an important role. The robot arm should have a realistic representation and it is useful in observing trajectories.

The main fields of application for manufacturing simulation systems are:

- planning and design,
- offline programming,
- testing and verification,
- supervising of the manufacturing process,
- quantitative measures for evaluating the system performance,
- training and education, and
- presentation (for marketing, the acquisition of financial support, etc.).

These activities shall be supported or taken over by the computer thus allowing the manufacturing design and verification to be done even before the first robot or NC machine is being installed at the factory plant.

However, simulation is not limited to proposed new systems or work cells; it can also be used to improve existing systems. For a manufacturing company with a limited budget and without a specific program to install robotics of advanced technologies, simulation can be used to test minor variations in existing systems and installing those that are cost-effective.

11.7 Three Steps to Correct Products and Tasks

Currently, the big share of robot / NC code is generated "by hand", either by programming the necessary activities and tasks manually or by using an offline programming system. As already stated, such a system allows a high-level task specification using a description language and / or taking the geometry information from CAD. This task specification is then transformed into standardized robot / NC code by means of computer.

The key question for correctness of production is the following: Having given a robot / NC code, applying it to the given raw stock using the mechanisms of a work cell by utilizing the specified tools, does the code lead to the desired workpiece, without doing harm to mechanisms, tools, and cell environment?

In [Schade and Schade, 1990], it is stated that (at that time) in the United States there were spent approximately 1.8 billion US $ per year for verification and correction of NC programs.

One reason for these high costs is that even for test runs the expensive NC machine equipment has to be used; additionally, the NC programs show a high error rate. Every test run of an NC program, no matter which material is used (wood, plastic, etc.), needs the same preparation time and manufacturing time as a production run.

The frequency of errors and the number of necessary test runs increases with the degree of production complexity. In the checked NC programs [Schade and Schade, 1990], about 10 % of drilling code and 20 % of turning code contained errors. This rate leapt to almost 50 % in case of 3D milling. There, nearly every second program contained errors that were so severe that the human expert on the NC machine had to return to the programming office in order to correct it. Complex machining operations, like 5D milling, showed an error rate of up to 95 %.

Based upon a suitable modeling of the objects within a cell and the task to be performed, the following strategies may be pursued:

1. The manually generated code is analyzed and checked for correctness by the aid of a computer. Subsequently, the code statements that cause errors have to be changed.
2. The robot / NC code is no longer generated manually but by the computer taking care of the correctness already during code generation.

Obviously, the former goal can be achieved more easily, since the degree of automation is lower.

In the following, we will briefly discuss these two strategies for achieving correct products and tasks. We also emphasize the importance of a comprehensive computer internal model in order to get a good representation of a manufacturing process.

11.7.1 Full Model Simulation

The first milestone for checking the correctness of a manufacturing task is the development of a suitable, homogeneous, computer internal representation (model) of the robot / NC world including the cell environment, mechanisms, tools, and workpieces. All necessary information shall be contained in this model, such that

- algorithms can be developed for simulating the motions of the mechanisms within the work cell as well as the impact of the tools (*simulation*), and
- algorithms can be developed for checking, whether the result of the robot / NC task satisfies the demands of the desired production process without doing any harm to the equipment of the work cell (*verification*).

For the structure of such a model, it is most important to be able to store geometric information as well as technological information in one interlinked data base. Furthermore, the model shall be designed in such a way that efficient algorithms for simulation and verification can be developed. Major attention has to be paid to the modeling of structures that change their geometric or technological properties dynamically (e. g., the raw stock, whose volume is steadily reduced by the machining process) and to the modeling of mechanisms (e. g., a robot or an NC machine). We call such a comprehensive representation of a robot / NC process *full modeling*.

11.7.2 Verification

Even when using an offline programming system, programming by "trial and error" would be too costly for practical applications, since test phases for every redesign of a work cell result in frequent standstills of the production (e. g., [Schade and Schade, 1990]).

In the simulation phase of an NC process, a geometric model of the workpiece is modified by the tool motions determined by the NC code. Verification of the correctness of the task requires a comparison of the geometric model of the workpiece with a geometric model of the desired part [Jerard *et al.*, 1989].

Furthermore, generally a work cell comprises objects that move when performing a task (mechanisms, workpieces) and objects that do not move (workbenches, pillars of the plant, desks, etc.). Certain objects are in contact with each other during the task (e. g., neighboring links of a robot, or the workpiece and the gripper) whereas others must not collide during the task to guarantee a correct (i. e. non-destructive) task (e. g., the gripper and a plant pillar, or the workpiece and the housing of an NC machine). If such collisions occur, much damage can be done to the objects involved.

In order to get a correct robot / NC program, the determined task errors must be corrected. If the errors can be traced to a specific gripper / tool motion, either manual or automatic correction of the cell geometry or the task file may be possible.

The following strategies can be pursued to detect errors of the manufacturing task:

- Human inspection during the animation process (*visual verification*):
  When pursuing this strategy, the user of the simulation system concentrates on the critical phases of the production task and tries to detect on the screen, whether the programmed task leads to non-machined areas, gouges, or collisions. This method necessitates a high resolution 3D graphic screen and much experience of the inspector.

  Visual verification is rather problematic, because the user might miss some of the critical situations when having to concentrate on the computer screen during the whole task. Even more important is the circumstance that there are errors that cannot be detected by visual inspection, because the on-screen animation is just a 2D projection of a 3D process. So, even if the screen resolution and the modeling quality (shaded solids, etc.) were improved, completely error-free tasks could never be guaranteed.

- Algorithms for detecting manufacturing errors (*algorithmic verification*):
  In a graphic manufacturing simulation system, the components of the cell are described in the computer by the cell geometry and the kinematics (and sometimes dynamics) of the mechanisms. The task that shall be performed by the mechanisms within the cell is described by the robot / NC code. Taking this full computer internal model, computer programs can, additionally to the simulation process, take over the verification of the tasks. For this method, fast algorithms for model comparison and collision determination are necessary.

We distinguish two different types of algorithmic verification, depending on different necessities for verification in a machining process and a manufacturing process:

1. The problem of automatic *machining verification* can be specified as follows:

   Given:
   - Geometry of raw stock.
   - Geometry of desired workpiece.
   - Tool geometry and tool motions (NC program).

   Decide:
   - When performing the specified machining operations, will the raw stock be transformed into the desired workpiece (with respect to a specified accuracy)? (If not, the errors in the NC program shall be pointed out and their impact on the workpiece – areas left non-machined or gouges – shall be visualized.)

2. The problem of automatic *manufacturing verification* can be specified as follows:

   Given:
   - Geometry of the work cell.
   - Kinematics of Mechanisms.
   - Robot / NC programs.

   Decide:
   - Can the robot / NC task be performed by the mechanisms in the work cell without violating technological restrictions like joint restrictions, or collisions caused by erroneous motions? (If the task specified by the NC code led to situations causing damage to the workpiece or one of the mechanisms, the NC statement(s) causing the error shall be detected and the location of the damage shall be visualized.)

According to [Rembold and Dillmann, 1986], algorithmic verification of robot / NC programs gives the following benefits:

1. optimum verification of a robot / NC task,
2. quicker implementation of a new robot / NC program,
3. reduction of tool breakage and of damage to the machining equipment, and
4. improved safety within the work cell.

In Chapter 12, we will further analyze the needs and possibilities of algorithmic verification, particularly concentrating on manufacturing verification.

11.7.3 Automatic Programming

The automatic generation of error-free robot / NC tasks would avoid all the problems with detecting and eliminating errors in the programmed tasks of the human expert. However, up to now, only code for a restricted class of – less complicated – robot / NC tasks can be generated automatically in a way that satisfies the technological and aesthetic demands of the human expert. Examples are:

- **Velocity profile planning**: Combining the geometric model with a model for the dynamic behavior of the machining process in order to determine optimum velocities and feeds of a given path automatically.

- **Task planning**: Automatic tool path can be currently generated only for "simple" NC machining processes, like sheet metal working, 2D turning, or 2D milling. For more complicated milling operations (3D, 5D), general approaches for automatic path generation are available only for ball-shaped tools, restricting the paths to zigzag milling.

- **Trajectory planning**: A future generation of substantially more autonomous and intelligent robots is expected to possess advanced capabilities of sensing, planning, and control, enabling them to gather knowledge about their environment, construct a symbolic world model of the environment, and use this model in planning and carrying out tasks set to them in high level style by an application programmer. [Latombe, 1991]

  Among these capabilities, planning involves the use of an environment model to carry out significant parts of a robot's activities automatically. The aim is to allow the robot's user to specify a desired activity in very high level, general terms, and then have the system fill in the missing low-level details. For example, the user might specify the end product of some assembly process, and ask the system to construct a sequence of assembly steps; or, at a less demanding level, to plan collision-free motions that pick up individual parts of an object to be assembled, transport them to their assembly position, and insert them into their proper places.

Although the full automation in the fields of robot / NC programming shall be kept in mind, a step-by-step approach, based upon simulation and verification, seems to be the best at the current stage.

11.8 Example: The KISS_ME System

11.8.1 A Concept for a Universal Graphic Simulation System

Typically, a modern company utilizes various CAD systems, NC machines, and robots. In most cases, different types from different manufacturers are used within one plant. Motivated by this development, factory automation converges to a new stage.

For this new generation in factory automation, software systems are necessary that, in addition to their original tasks, are capable of combining, structuring, and organizing the already installed hardware and software of various manufacturing devices. Such *universal systems* should not only be responsible for simulation, but also be able to exchange data with different CAD systems via standardized interfaces. Furthermore, several devices from different manufacturers – such as robots, NC machines, transportation systems, feeding or removal units – must be supported. This fact necessitates universal modelers and standardized input / output interfaces.

11.8.2 KISS_ME - A Universal System for Graphic Simulation

KISS_ME (Kinematic Interactive Simulation System for Mechanical Engineering; [Peneder, 1996]) has been designed as a flexible turnkey environment for graphic simulation and programming of robots and NC machines, being based upon one of the earliest commercially available manufacturing simulation systems, SMART ([Mayr *et al.*, 1989], [Mayr and Öllinger, 1991]). KISS_ME meets all major demands of the shop floor. It supports factory design and planning in the following way (Fig. 11.3):

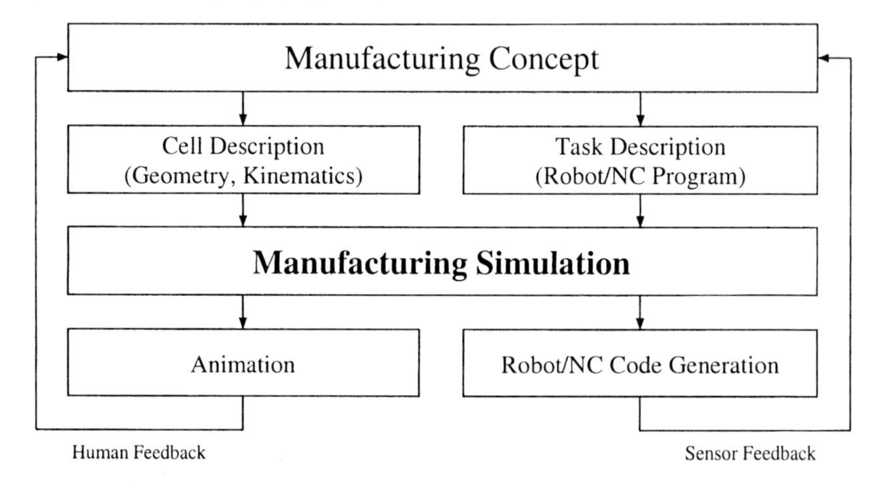

Figure 11.3: The structural concept of KISS_ME

Starting with the manufacturing concept, the planning stage splits up into two parts,

1. modeling the manufacturing cell and
2. programming the manufacturing task in a high-level language.

The first part can either be done using any CAD system that is able to pass output data via a geometry interface (IGES or PDES / STEP) or by including object data bases (containing predefined descriptions of robots, pallets, NC machines etc.). In that way, the design of a cell is reduced to choosing suitable mechanisms and objects. Thus, the time necessary for repeated design activities is saved. In KISS_ME, the production cell is stored in an object oriented "world model" data base using a CSG representation that is restricted to the sole use of the union operator. Such a representation can easily be gained from CAD systems. It needs a small amount of storage and is well-suited for fast collision checks due to the hierarchical modeling structure. For the coding of production tasks, KISS_ME offers *RobLan* (see [Mayr and Oberreiter, 1986]), a general, task oriented programming language. RobLan allows symbolic programming of arbitrary manufacturing devices using the world model data base.

The simulator of KISS_ME supports control, analysis, and verification of an arbitrary manufacturing process via the device-independent animation system. Any modern graphics hardware can be used via the graphics interface (supporting PHIGS / PEX and OpenGL). Using the solid object representation, a realistic 3D animation is possible.

When the animated process satisfies the intentions of the designer, KISS_ME automatically produces device-independent control data for all mechanisms involved. Device-adapted post-processors transform these data into suitable control sequences for the actual machines and robots.

The visual information gained by the animation system can be used for revealing discrepancies between the intended and the programmed production process (feedback). Thus, the manufacturing program can be corrected already during the design stage rather than on the shop floor. Furthermore, algorithmic verification (see Ch. 12) and sensor information (see Ch. 13) from a first test run can be used, e. g., for calibration purposes.

11.8.3 Object-oriented Features of KISS_ME

Today enormous technological development, especially in computer science, requires to re-design and extend application software frequently. This is one of the reasons for the development of the KISS_ME class library in a pure object-oriented way. This technique opens the possibility to make a hierarchical and modular structure of software packages.

The abilities of KISS_ME last from designing a work cell with one robot up to control a whole virtual factory. Through the concept of interaction, based on the graphic user interface, one can control the whole scenario and change several parameters immediately without reprogramming the whole work cell. KISS_ME brings together all these points. The modular concept allows to extend KISS_ME into any direction (like, e. g., architectural design).

11.8.4 The Universality of KISS_ME

KISS_ME fulfills all requirements of a universal simulation system as specified in Ch. 11.8.1. This system can not only be used for simulation purposes, but also for the exchange of data with different CAD systems via standardized interfaces. Furthermore, arbitrary production devices from different manufacturers – such as robots, NC machines, transportation systems, feeding or removal units – can be simulated and programmed by a universal modeling concept and standardized input / output interfaces. A simple scenario modeled with KISS_ME is shown in Fig. 11.4.

11.8.5 The KISS_ME Algorithm Test Shell

A test shell included in the KISS_ME system allows the mounting of various routines for collision detection and collision avoidance methods into KISS_ME, see Fig. 11.5. This facilitates efficient testing and comparison of new algorithms for task verification.

In particular, we classify algorithms for task verification and task generation in the following way, according to the problems they solve:

1. **Algorithms for collision detection:** Having given a specified situation within the work cell (characterized by the simulation time), such algorithms decide, whether there is a collision between objects within the work cell.

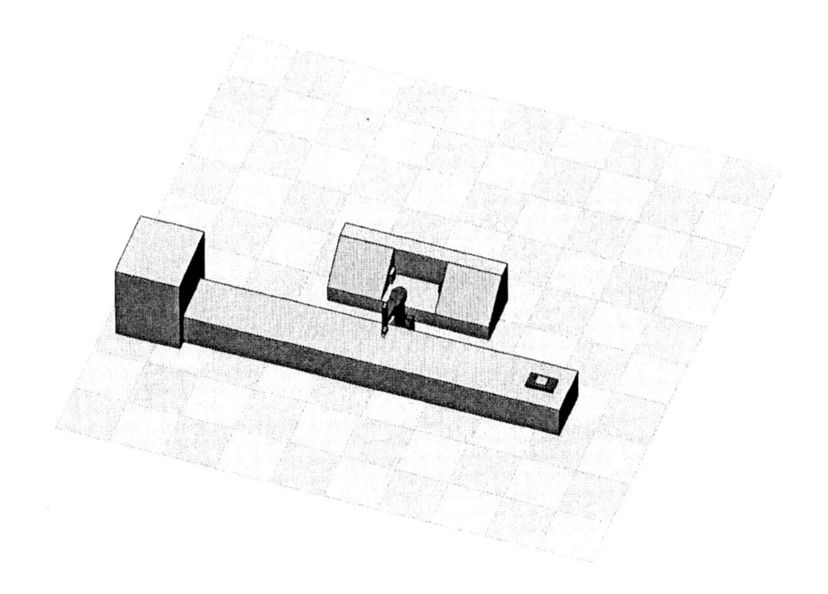

Figure 11.4: Simple scenario modeled with KISS_ME

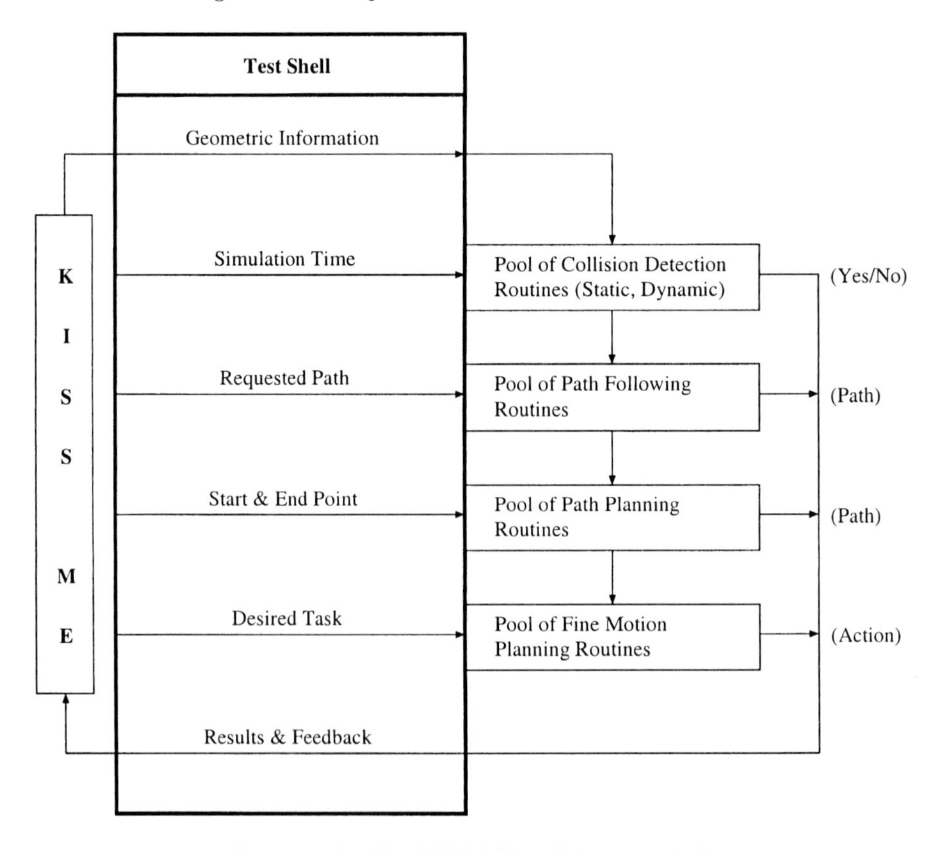

Figure 11.5: The KISS_ME collision test shell

2. **Algorithms for path following:** A straight line (in the Cartesian coordinate system) is not necessarily the most efficient path to be followed by a robot since continuous path adjustments reduce the moving velocity enormously. Path following algorithms take a Cartesian path (consisting of translational and rotational motions) as input and construct a path that does not deviate more than a specified accuracy from the given path but allow a faster and smoother motion of the robot.

3. **Algorithms for path planning:** Such algorithms take the start point and the end point of a desired motion as input and create a collision-free path for moving the mechanism according to the specified locations. Furthermore, the resulting paths have to be further adapted (smoothed, adjusted to acceleration and velocity limits, etc.) in order to be applicable in practical use.

4. **Algorithms for fine motion planning:** Utilizing the geometric information about the work cell, such algorithms determine control sequences for smooth grasping and positioning of objects, balancing, etc.

Chapter 12

Algorithmic Manufacturing Verification

This chapter gives a survey on algorithmic manufacturing verification, particularly on determining collisions in a manufacturing cell. Different collision checking problems are stated and a survey on solutions to these problems is presented.

12.1 Goals of Verification

The following types of constraints have to be checked when verifying a manufacturing task:

- **Technological constraints** are either feasible characteristics for the parameters of the objects involved in the task (e. g., velocity profiles for the joints of the mechanisms[1]) or certain dependencies between the objects (e. g., the weight of the pay load that may influence the workspace where a robot can work sufficiently exactly; or the type of material of a milling tool that influences the feed rate).

  In most cases, parameter characteristics are already considered when designing the task [Shimada *et al.*, 1989]. For object dependencies, table-look-ups and data bases (often called "expert systems", "knowledge bases", and the like) are adjoined to simulation and programming systems [Weber and Dürr, 1990].

- **Logical constraints** are restrictions due to chronological, judicial, economic, etc. dependencies. They constitute, e. g., constraints imposed by laws or standards, rules found by artificial intelligence methods, or restrictions by the synchronization of parallel tasks. They are treated in a way similar to technological dependencies [Krautter and Steinert, 1989]. A rather new approach to include logical constraints into design and programming is the use of constraint logic programming (CLP) languages (see [Cohen, 1990] for a survey). This scheme defines a class of languages designed to reason about constraints using a logic programming approach that bases upon a rather general operation called constraint satisfaction. Applications to engineering problems are, for instance, sketched in [Heintze *et al.*, 1988].

[1]When talking about mechanisms, we include industrial robots, handling devices, and NC machines. In accordance with the literature, sometimes the term "robot" is used as a synonym for "mechanism".

- **Geometric constraints** are restrictions of the mobility of the objects. They are caused either by limits for the joints of the mechanisms involved or by obstacles that restrict the task space. If a moving object inadvertently touches or crashes into a stationary object or two moving objects hit each other, this situation is generally called a *collision* between objects.

Checking for geometric constraints shows different levels of difficulty, depending on the type of the checked constraints:

- It is rather straightforward to check, whether the joints of the mechanisms are at a feasible position or not. If the joints are independent from each other, it has to be simply checked for each of the joints, whether its current value is within the specified interval(s). If dependencies do occur, the interval limits can be parameterized.
- The task of checking, whether a collision does occur or not, however, is much more complicated. More than thirty years have passed now, since this question was stated first in connection with robotics / NC [Pieper, 1968]. However, the ultimate general solution has not been found till now. More probably specific approaches for different situations and problem variations will tackle the problem better.

In the following, we will focus on the collision problem being one of the hard nuts to crack in algorithmic task verification. We will structure the problem hand in hand with giving a survey on known approaches to attack the collision problem. In the sequel, we specify the collision checking problems and their respective modeling requirements.

12.2 Verification of Manufacturing Tasks

Basically, one distinguishes two classes of verification of manufacturing tasks:

1. In *cell status verification*, one has given a snapshot situation of the considered cell and wants to know whether certain / any objects within the cell do intersect with others or not. Since this kind of verification does not include any motions of mechanisms, the corresponding problem is called *static collision problem*.
2. In *algorithmic task verification*, not only the geometry and the position of the objects within a work cell is considered, but also the task performed by the mechanisms. This means that one wants to check the whole process for collision.

Obviously, algorithms for attacking the task verification problem must be more powerful than algorithms for cell status verification. This fact is also reflected in the literature. Whereas there is already a number of applicable methods for cell status verification (see, e. g., [Myers, 1981]), algorithms for task verification (this problem is generally called *dynamic collision problem*, since motion is included) are still rare.

12.3 Collision Checking

Collision checking can be done directly on the mechanism ("online") or using a computer and a simulation system ("offline"), see Fig. 12.1. Whereas for online checking sensors play the key role and no or just a simple mathematical model may be necessary (depending on the sensor types), a detailed model for both the mechanism(s) and the environment is essential for offline checking.

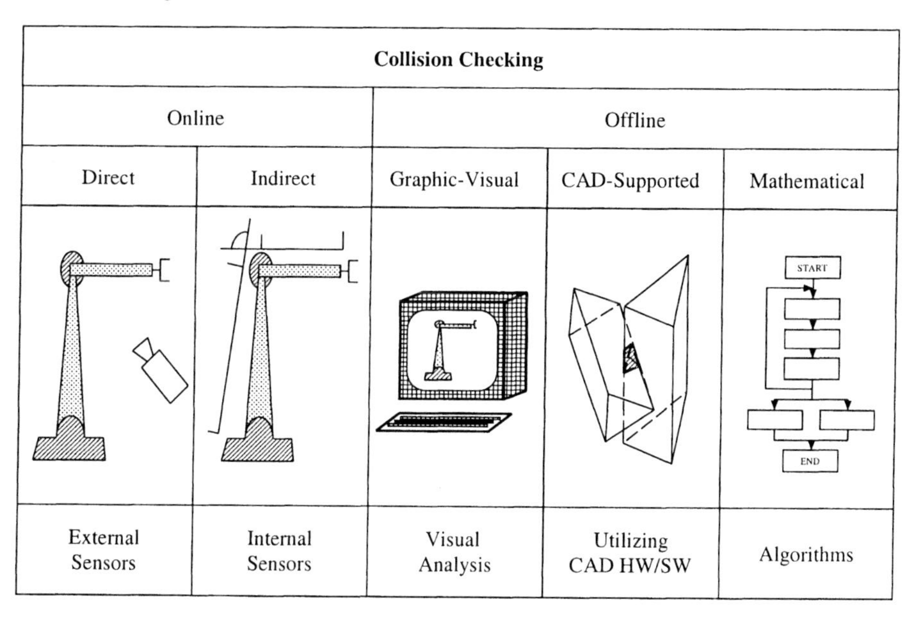

Figure 12.1: Different methods for collision checking

12.3.1 Online Collision Checking

Online collision checking is done via sensors that are placed directly on the moving mechanisms or in the environment. Collisions can be detected either in the moment of contact (by means of tactile sensors) or in advance (by means of distance sensors). For details on different strategies and types of sensors, we refer to, e. g., [Desoyer *et al.*, 1985]. When a collision is detected, the reaction on it can be different depending on the capabilities of the sensors and the control system:

- When a plain system is used, the mechanisms may be "simply" stopped (*emergency stop*). (Simply is not quite correct, because just switching the power off does not work for a robot or an NC machine; an immediate halt in the current or next possible defined position is desired.)

- When a more sophisticated system is used that includes a model of the environment, a correction of the motion may be possible in order to get a safe trajectory.

The main advantage of online collision checking is that it is possible to determine dangers of collision that are generally unforeseeable (e. g., parts dropped by a malfunctioning crane, non-planned motions of other mechanisms), if suitable sensors are on the respective positions. For the same reason, online collision checking plays also an important role for the inclusion of human workers into a manufacturing cell, because their actions and motions can never be planned exactly.

The disadvantage of mere online checking is that design errors cannot be detected until the first test run at the shop floor. No support during the planning stage can be given. Also, the elimination of collision situations necessitates frequent human actions in the work cell and increases the time that is necessary for reprogramming a work cell.

Two methods for online collision checking can be distinguished depending on the way sensors are used (Fig. 12.1):

1. **Direct online checking**: Sensors are used that are external to the robot and observe the robot's motions. According to [Duelen *et al.*, 1990], the main sensor types used are tactile, acoustic, and optical sensors. No model for the robot is necessary. The real world itself is taken as the environment "model". Only an emergency stop is possible as a reaction to a detected or foreseen collision.

2. **Indirect online checking:** This method uses a simple computer model of the robot that allows the assignment of trajectories and uses robot-internal sensors for measuring deviations from these trajectories. Whereas the geometry of the robot can be gained from this model, the actual positions of the links are determined in real time by suitable sensors. Using fast computer hardware, collisions can be determined in real time, i. e. at least as fast as the robot controllers and servos are able to move the robot.

   By modeling selected points in the environment and using external sensors for determining the environment of a possible collision, a correction of the motion that would lead to a collision is made possible yielding a collision free trajectory [Lumelsky, 1986].

12.3.2 Offline Collision Checking

The main applications of offline collision checking are the following:

- determination of the *reach* and the *workspace* of a given mechanism,
- *collision detection*, i. e. predicting, whether a particular action would result in a collision [Udupa, 1976],
- *sensor modeling* (simulating touch sensors, approach sensors, distance sensors, etc.),
- *collision avoidance*, i. e. what to do with a moved object (robot arm, tool, etc.) once a collision on the original path has been forecast [Myers, 1981],
- *motion planning* and *fine motion planning* (grasping, positioning, assembly, etc.).

These applications result in the following requirements for a universal collision check:

- universal applicability to all kinds of mechanisms and cell situations,
- complete (at least conservative) checking, i. e. a "no collision" response must be definitive,
- automatic test run that needs no human interaction,
- low computation times to allow interactive simulation.

According to Fig. 12.1, the different approaches to offline collision checking can be classified in the following way:

1. **Visual analysis**: If the process is just simulated on a display and the verification is left to the user (by means of visual inspection, see, e. g., [Pinkler and Simon, 1976]), the task description or the cell layout must be altered by repeated manual interventions until collisions are eliminated [Anderson, 1978]. Moreover, in some cases visual inspection cannot give complete security since today's output devices only offer 2D views of a 3D process. Note that this is not really a collision test, since the analysis has to be done by the user.
2. **Utilizing CAD hardware or software**: Certain features of a CAD package or even the hardware system are used to take a region in space (representing the object to be tested for collision), and discard those portions of the work cell that lie outside the region. The scene is then displayed, with all the objects lying outside the bounding box being discarded. If any portion of the scene remains, there was an intersection between the tested object and the environment. If nothing remains in the bounding box, there was no intersection. [Hoffman and Hebert, 1986] use hardware (a "geometry engine") for performing this operation, [Diedenhoven, 1985] uses the facilities of a solid modeler. [Smith, 1985a] and [Smith, 1985b] facilitate the clipping process implemented in specially designed VLSI chips.
3. **Mathematical-algorithmic approaches**: Algorithms are used that have been particularly designed for the purpose of collision checking. In contrast to solid modelers, these algorithms often do not generate the intersection volume, if the checked objects do collide. They only determine one point that is used as a representative of this volume. On the other hand, in several approaches "witnesses" are determined that certify the disjointness of the checked objects. Such algorithms have been developed for stationary situations ("static" checks) as well as for situations including moving objects ("dynamic" checks). In the following, we will concentrate on classifying and discussing various such algorithms.

12.4 Offline Collision Problems

In [Weck and Stöck, 1985], the term *collision* is defined to be the inadvertent clash of two bodies in time and space, where at the moment of the clash at least one of the objects has a kinetic energy that is different from zero. If an intended contact does occur between a gripper or a tool and the workpiece for the reason of handling or machining, this is not called a collision [Schöling and Reles, 1983]. Fig. 12.2 illustrates the information flow and data flow that is necessary for collision checking.

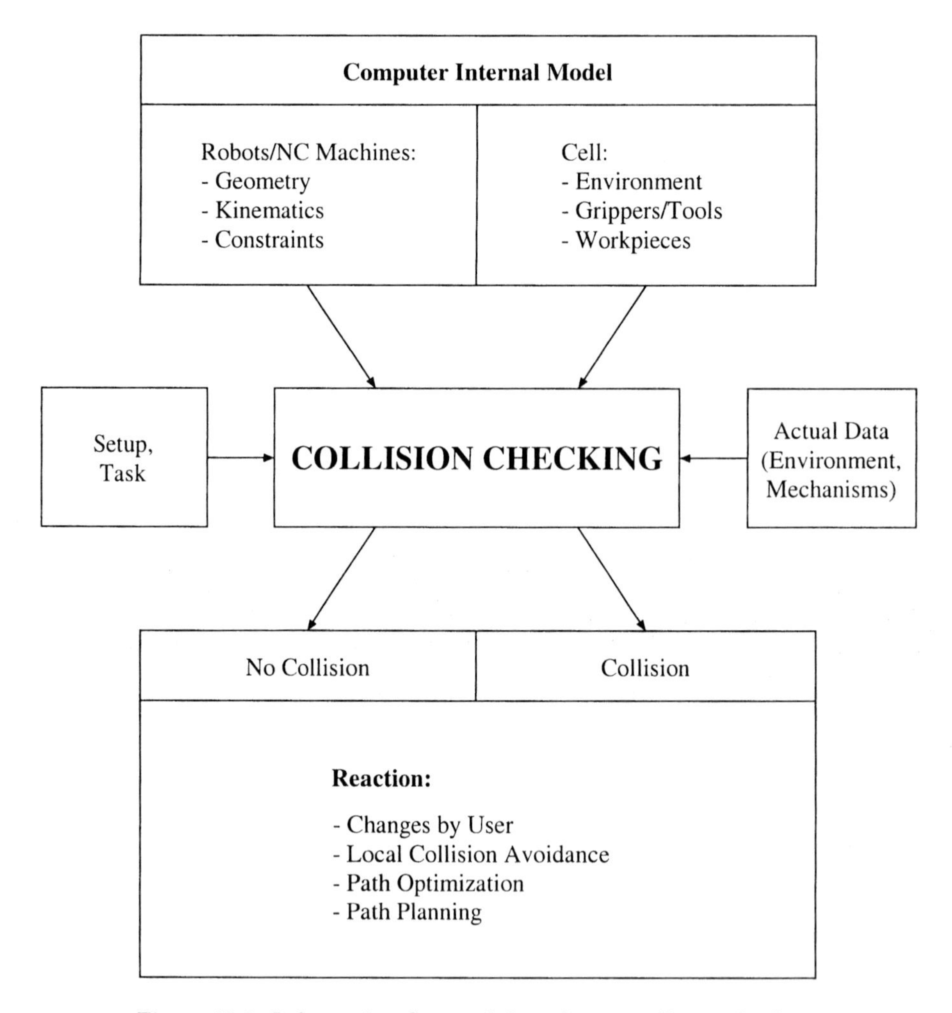

Figure 12.2: Information flow and data flow in collision checking

12.4.1 The Static Collision Problem

Static collision checking deals with situations where one has given a snapshot of the considered cell and wants to know, whether certain / any objects within the cell do intersect with others or not.

Such situations might for instance occur when:

- One stops the simulation at a critical point, because from the visual impression one may suppose that there might be a collision at exactly that moment. Now one wants to have this suspicion proved or disproved.
- One designs a work cell and wants to get some aid for the placement of objects. For instance, one wants to place a feeding unit at a certain position, but does not know, whether it can be reached by the robot or not.

Since this kind of verification does not include any motions of mechanisms, the corresponding problem is called *static collision problem* [Myers, 1981].

The first algorithmic idea for this problem could be (let the snapshot situation be characterized by the simulation being stopped at time $= t_0$):

1. Determine the positions of all objects involved at time $= t_0$.
2. Take all pairs of objects[2] that should not collide at time $= t_0$.
3. Find out, whether they do collide or not.

Mathematically, this problem can be formulated in the following way:

Problem P-S "Static Collision Problem":

Given: P_1, P_2. `-- Objects`

Decide: $P_1 \cap P_2 \neq \emptyset$? `-- ''Intersection''`

This strategy, also-called "stepped move approach" in [Myers, 1981], is illustrated in Fig. 12.3.

Soon after the first "Eureka!" one will begin to ask, "O. K., now I know about the situation at $t_0, t_1, t_2 \ldots$, but what if there is a collision, say, between t_0 and t_1?". Due to the time-discrete checks, there is the possibility that the system might miss a glancing collision, because the actual collision would occur between stepped moves [Myers, 1985]. There are two methods of ensuring against such an occurrence. The first is to enlarge the representations of object volumes by the length of the path that the arm moves between two snapshot scenes. This conservative approach will guarantee that no actual collisions pass undetected by the system,

[2] Collision detection means to determine whether any two objects in a set do intersect. Intersection algorithms for sets of objects, however, usually test, whether the common intersection of all objects is non-empty. For the modeling consequences of this circumstance and alternative approaches, see Ch. 12.5.1.

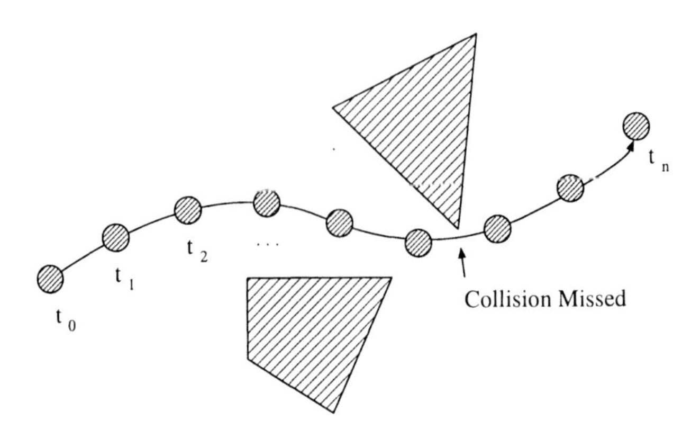

Figure 12.3: Static collision checking

but may result in false alarms, predicting a collision when none would occur. The second method is to shrink the step size, which is under control of the user. This procedure decreases the difference between consecutive scenes in a stepped move, which means that more scenes have to be checked for a given motion. Since the time to evaluate a scene is approximately constant, accuracy is gained at a cost in response time.

A better alternative is to get more information out of a single check, like "If I know that there is no collision in t_0, can I derive any information about the near future, e. g., a moving distance limit that guarantees a collision free motion within this distance? Or, can I guarantee collision-freeness between two static checks by some theorem?"

This *dynamic forecast* into the near future can be formulated in the following way:

Problem P-SD "Static Collision Problem with Dynamic Forecast":

Given: P_1, P_2, `--Objects`
$[t_S, t_E]$, `--Maximum time interval to be checked`
$M_{t,1}, M_{t,2}$. `--Motion matrices describing the transformation of` P_1 `and` P_2 `at time` t, $t \in [t_S, t_E]$

Find: t, such that `--Time offset`

$$t = \max\{t' | t_S \le t' \le t_E, (\forall t_S \le t'' \le t')(M_{t'',1}P_1 \cap M_{t'',2}P_2 = \emptyset)\}.$$

If no such t exists, report this fact.

Both rotations and translations can be considered defining the transformation matrices in a suitable way. Often, it may be sufficient to compute a t that is "large enough", instead of computing the actual maximum value at a large time expense.

If only motions at constant velocities are considered, for each t a corresponding motion offset d can be derived that defines a certain environment, where the two objects may move without any danger of collision. Both the translational and the rotational components have to be considered when determining such a d.

With an algorithm solving this problem one can now start with one test, get a time limit t

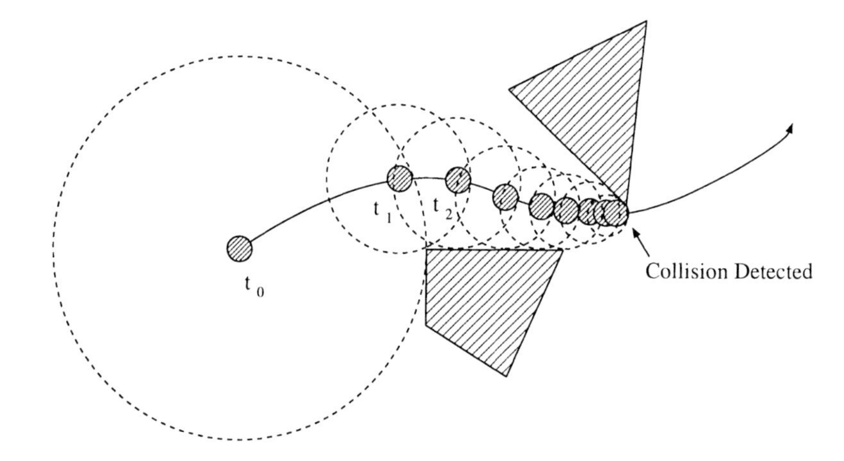

Figure 12.4: Static collision checking with dynamic forecast

or a distance limit d, and continue with the simulation until one object has moved till to the limit, Then the next test is performed, taking the time limit of the previous test as the start time of the new interval, and so on. The points where there is checked for collision are in this way adapted to the environment within the work cell (Fig. 12.4). This procedure guarantees the detection of any collision during the whole task. However, practical tests have shown the following problem: In critical situations (i. e. small distances), naturally, tests have to be performed more frequently. Under certain circumstances, in such situations the distance limit (and consequently the time until the next test) converges to 0, and the algorithm does not stop. Here, only a lower bound for the distance limit guarantees the termination of the process – but no longer a collision free task. A better alternative is to reason from static checks at two positions about possible collisions along the path between them.

12.4.2 The Dynamic Collision Problem

A different approach, avoiding the problems occurring in static collision checking, is to not only consider the geometry and the position of the objects within a work cell, but also the task performed by the mechanisms. This means that the whole task is checked for collision. Clearly, the algorithmic complexity of such an approach to *dynamic collision checking* is higher than that of static collision checking.

Situations that can be checked in this way are:

- one object that moves along a trajectory passing-by a stationary object (e. g., the gripper of a robot vs. supplementary devices), and
- two objects that move along trajectories (parallel tasks within a work cell).

These situations cover all cases of motions to check that may occur.

Consequently, one ends with the question, "Given a time interval [0, t], is there any collision within this interval?". The corresponding problem is generally called *dynamic collision problem*, because the motions of mechanisms are included [Myers, 1981]. This problem can be formulated in the following way:

Problem P-D "Dynamic Collision Problem":

Given: P_1, P_2, --Objects
$[t_S, t_E]$, --Time interval to be checked
$M_{t,1}, M_{t,2}$. --Motion matrices describing the transformation of P_1 and P_2 at time t, $t \in [t_S, t_E]$

Find: t, such that --Time of first collision
$t = \min\{t' | t_S \leq t' \leq t_E, M_{t',1}P_1 \cap M_{t',2}P_2 \neq \emptyset\}$.
If no such t exists, report this fact.

In the general case, the whole time interval $[t_S, t_E]$ is tested by one dynamic collision check. In order to improve the algorithmic performance, special algorithms try to split the motion (and, thus, the time interval) into a finite number of simpler pieces. A possible strategy is to split a motion into sub-trajectories that can each be described by a single transformation matrix. Such a strategy for solving Problem P-D guarantees the detection of any collision occurring during the task by a finite number of checks (Fig. 12.5). This number is generally very small, particularly when the path is rather linear. Moreover, the user can guarantee that the time intervals do not converge to zero.

Strictly speaking, only algorithms solving the dynamic collision problem can be called "collision checks", because only they can guarantee to determine any collision during a manufacturing task.

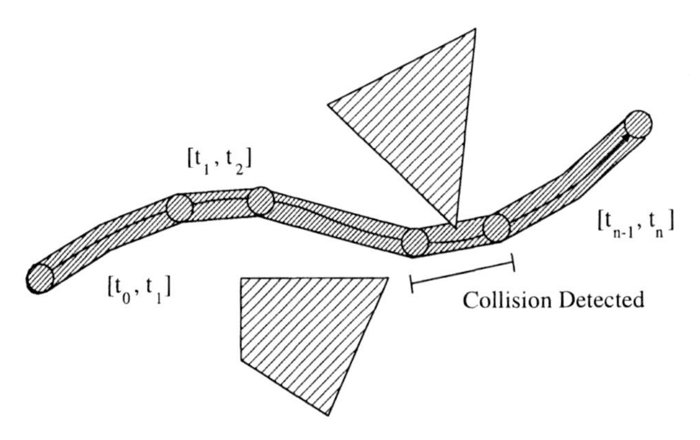

Figure 12.5: Dynamic collision checking

12.5 Solutions to the Collision Problems

12.5.1 Modeling Types and Paradigms

The following modeling paradigms are generally assumed for the collision checking algorithms:

- **Convex objects**: The majority of the approaches supports only the detection of collisions between convex objects. Non-convex objects, therefore, have to be split up into convex ones in such approaches.
- **Conservative enclosures**: If the objects are modeled in too much detail, a simplified approximation of their shape is used for collision checking. It is important to not just approximate the objects by volumes of simpler topology, but to really enclose them by the volumes. Only in this way a conservative behavior of the collision checking algorithms can be guaranteed (i. e. if the enclosures do not intersect, the original objects cannot intersect, too), see Fig. 12.6.

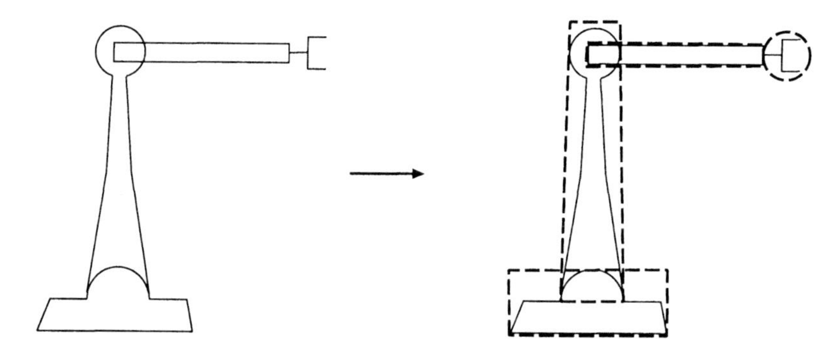

Figure 12.6: Enclosing objects for conservative checking

Several approaches (e. g., [Myers and Agin, 1983], [Potthast *et al.*, 1988]) perform this enclosing strategy on several levels to enable a hierarchical checking strategy. We refer to [Mayr, 1991b] for details.

- **Checking in pairs**: In order to be able to decompose collision detection into a number of collision checks for pairs of primitives, all volumes must be of positive extent (i. e. no volume must represent a hole). Therefore, only the "union" boolean operator is permitted for composing objects from simpler ones [Myers, 1985].

  There are a few approaches that, instead of checking in pairs, check for a whole configuration at once, whether any two objects do intersect: In a study of the 2D case, [Sharir, 1985] developed a generalization of Voronoi diagrams [Voronoi, 1908] for a set of (possibly) intersecting circles, and used this diagram for detecting intersections and computing shortest distances between discs. [Hopcroft *et al.*, 1983] present an algorithm for detecting intersections between a number of 3D spheres.

12.5.2 Algorithmic Strategies

12.5.2.1 2D Collision Checks

There exist several approaches for collision detection in 2D (e. g., [Balbach and Sollmann, 1982], [Pilland, 1988]). Such approaches find their applications mainly in turning, where in many cases it is sufficient to describe the machining process in 2D only. Because of their restricted field of applications and the relatively simple basic ideas, 2D collision checking methods are not discussed further here.

12.5.2.2 3D Static Checks

Mainly, the following different approaches can be found in the literature:

- **Intersecting polygonal facets**: Several approaches enclose the objects by polytopes and store the topological information of these polytopes (e. g., [Baumgart, 1974], [Stobart, 1987]). The facets of the polytopes are then checked for intersection in pairs. Since each of the facets of the first polytope has to be checked against all of the facets of the second polytope and vice versa, the number of necessary intersection checks grows quadratically with the modeling detail of the objects. The advantage of this method is that it allows a large number of types of objects to be modeled exactly, including non-convex polytopes.

- **Intersecting cylinders and spheres**: Due to their topology, cylinders and spheres can be checked very efficiently, whether they do intersect each other or not, particularly when the cylinders are thought of being stretched to infinite lengths. Consequently, these primitives were frequently used in the beginning of algorithmic collision checking (e. g., [Pieper, 1968], [Khatib and LeMaitre, 1978]). However, for many objects it is difficult to describe them sufficiently exact by mere cylinders and spheres. As a consequence, [Kondo and Kimura, 1988], for instance, enclose the objects hierarchically by an increasing number of spheres. For describing a link of a mechanism whose shape is not close to a ball, several adjacent spheres of same size are being used.

- **Intersection algorithms between primitives of different kinds**: [Myers, 1985] gives a survey on various intersection algorithms between primitives like spheres, boxes, cylinders, infinite planes, polyhedra, ellipses, and surface patches. [Stobart, 1987] concentrates on intersecting boxes and cylinders, [Schöling and Reles, 1983] on cones and pyramids. Many of the intersection algorithms base upon the methods presented in [Boyse, 1979]. Numerical problems are reported frequently (e. g., [Lozano-Perez, 1977], [Cameron, 1984]). [Schmiedmayer, 1987] uses only boxes and regular prisms as primitives, but gives several hints to speed up the single checks.

  To illustrate that efficiency can be gained when looking in more detail onto a collision checking routine, no matter how simple it may be, we cite [Myers, 1981]: "Sphere-sphere intersection checking is the fastest of all intersection algorithms. Normally, spheres are tested for intersection by computing the square of the distance between the centers of the two spheres, and comparing this against the square of the sum of their radii. However, this requires four (floating-point) multiplications. Our system first checks the unsigned differences of the coordinates of the centers, along the x, y, and z axes. If any one of these differences is greater than the sum of the radii, then the spheres cannot be in intersection. Next, we check the sum of the differences. If this sum is less than the sum of the radii, then the spheres are intersecting. Both of these tests require only addition of numbers, and are therefore significantly faster than testing the distance directly. Finally, if both of these tests are inconclusive, the square of the distance is found and compared with the square of the sum of the radii".

- **Using rectangular parallelepipeds**: For a large class of "industrial" objects rectangular parallelepipeds – and often even isothetic ones – give an approximation of the scene of sufficient quality. [Hui and Yi, 1988] pursue this approach and represent cubes by intersecting half-spaces. [Yu and Khalil, 1986] use special tables to reduce the number of object pairs that have to be tested by discretizing the environment and tracing the motions of the mechanisms. [Potthast *et al.*, 1988] restrict their enclosures to isothetic boxes to yield faster intersection checks. A different approach to check whether two boxes do intersect is pursued in [Ozaki *et al.*, 1984]. They use the centers of gravity of the boxes together with their position frames to decide this question.

- **Spatial enumeration techniques**: This method tries to approximate the objects by a number of very simple volume primitives and checks for intersection by determining, whether the two sets of volume primitives are disjoint or not. Even though the number of volume primitives in the sets is generally very high, the test for set disjointness is still often rather fast, because the single checks for set containment are generally very simple. A very popular approximation method is the use of octrees, where the volume of the object is partitioned into isothetic cubes [Ahuja *et al.*, 1980]. [Chen *et al.*, 1984] describe how three-dimensional objects can be approximated by a set of hexagonal cylinders that are symmetrical with respect to the z-axis and represent these cylinders by means of septrees. [Anderson, 1978] uses 3D "histograms" and compares the local heights of this diagrams to check for intersection with a milling tool. A combination of octrees and run-length encoding is used in [Mayr and Heinzelreiter, 1991b] for the modeling of workpieces and the detection of collisions during machining operations. The disadvantages of spatial enumeration methods are that they tend to be very memory consumptive and that the model in most cases has to be regenerated when an object is rotated.

- **Superquadrics modeling**: [Dombre *et al.*, 1985] use superquadric (superellipsoidal, supertoroidal, and superhyperboloidal) primitives for collision detection. Mathematically, superquadrics are closed surfaces generated from the spherical product of two 2D curves [Barr, 1981]. They provide a powerful tool to build solids with a restricted number of parameters. A large variety of shapes can be constructed from a single equation by altering a few interactive parameters. The properties of superellipsoids that are important for interference checking (mainly inside-outside functions) are listed in [Comba, 1968]. Intersection algorithms for superellipsoids with a number of other objects are stated in [Stifter, 1986]. [Baldur and Dube, 1987] discuss the intersection of ellipsoids that are symmetrical with respect to a coordinate axis. Intersection checking based upon superquadrics is in the range of seconds on a standard workstation today.

- **Determining a witness for disjointness**: In [Roider and Stifter, 1987], a method is described that tests two arbitrary convex objects for intersection. The idea is that for each two convex objects in the plane that do not intersect there exists a pair of straight lines such that one object is totally included in one wedge with which the second object does not have a point in common. In 3D space the straight lines are replaced by three-faced pyramids. [Roider and Stifter, 1987] describes an iterative algorithm that finds such a wedge if the given objects do not intersect ("witness for disjointness"). The algorithm returns a point in common to both objects if they intersect. Other types of witnesses are, e. g., separating surfaces or bundles of rays attesting disjointness. Details on these strategies are presented in the next two items.

- **Minimal distance computation**: Minimal distances between polyhedral objects are used in the approaches of [Pritschow and Kayser, 1987] and [Stobart, 1987] for determining, whether they do intersect or not. However, the determination of minimal distances turns out to be rather slow, particularly when complex objects are involved or the objects are being composed of a larger number of primitives. [Red, 1983] gives a method that is quite efficient for 2D polygons, but it is stated there that they cannot specify a method that is similarly efficient in three dimensions.

- **Ray casting methods**: This approach uses numerical ray tracing methods similar to those used for generating pictures of solids (e. g., [Cordonnier *et al.*, 1985]), and typically takes several seconds to test a single position. The number of rays that is necessary and sufficient for drawing conclusions about the intersection of two objects together with their arrangement is discussed, e. g., in [Cole and Yap, 1985].

12.5.2.3 3D Dynamic Checks

Mainly, the following different approaches can be found in the literature:

- **Swept volume approach**: In the swept volume approach (the term is due to [Widdoes, 1974]), an attempt is made to represent the volume swept out by the moved object, then to intersect this virtual object with the objects in the scene [Myers and Agin, 1983]. This method is pursued, e. g., in [Liegeois *et al.*, 1984] and in [Warnecke and Altenhein, 1986], the latter using particularly prisms as basic objects. Often, the motion can be either purely translational or rotational about a single axis [Lozano-Perez, 1980a], [Lozano-Perez, 1980b], [Brooks, 1982].

  Generally, there may be only one moving object in this approach. If more than one object is moving, then pairs of objects must be analyzed, considering in each case their relative velocity and motion. This can lead to complicated paths and large computational expense [Stobart, 1987].

- **Configuration space techniques**: If the swept volume approach is combined with the technique of shrinking the representation of the moved object and growing the obstacles by an equivalent amount (i. e. determining the "configuration space" of the scenario; for an exact definition of the configuration space, see [Lozano-Perez, 1980b]), the swept volume reduces to a curve in space [Lozano-Perez and Wesley, 1979]. Again, only one object may move to guarantee paths that are not too complicated.

  A common way of object modeling is to enclose the links of the robot by cylinders, and then shrink them accordingly. [Udupa, 1976] models the obstacles by polyhedra and intersects them with the "line robot". The author notes that his approach works only well for a simplified 2D model. In [Schöling and Reles, 1983], distance computations are made between the trajectory line and the obstacles to check for collisions. This method permits a larger class of obstacle descriptions, but it also implies a higher number of computations compared with line intersection. A similar approach is pursued in [Gerke, 1985]. There, instead of computing distances, angles between axes are compared in order to check for collision.

- **Computing bounding boxes**: In [Potthast *et al.*, 1988], a very rough check for moving objects is suggested: Using their method, a bounding box is generated for each of the objects involved. The bounding box of the moved object is now swept along the trajectory and the bounding box of this swept volume determined. This new box is used for intersection checks.

- **Reasoning using static checks**: In [Stifter, 1989], the information gained from static collision checks is used to derive information about collisions during a whole motion. This method first generates witnesses for disjointness at certain positions along a specified path of a moving object with respect to stationary objects. Then these witnesses for disjointness are used for reasoning about the collision-freeness of the whole path. Theorems are stated, when this collision check can be applied (e. g., due to restrictions to certain types of paths) and which basic operations are necessary in order to be able to derive conditions that guarantee that a whole motion is collision-free.

- **Four dimensional extruded solids**: This method considers the time during which the motion takes place as a fourth dimension [Stobart, 1987]. The three-dimensional solids making up the moving objects are "extruded" into four dimensional space and analyzed using extended geometric facilities. Collisions are detected by looking for common volumes between these four dimensional solids. A problem is that the intersection represents the entire history of the collision and it may be difficult under certain circumstances to detect the moment of the first collision. Extruded solids are used, e. g., in [Cameron, 1984] and [Cameron, 1985] to determine collisions between polygonal objects undergoing piecemeal linear motions. There, objects are modeled as a set combination of half-spaces. For determining intersections, a null-object detection routine is used. The computation times are in the range of seconds due to the iterative character of the routine used for null-object detection.

- **Algebraic approaches**: [Moore, 1980] uses interval methods to solve a system of nonlinear equations. Although objects can be described by such equation systems quite accurately, the used method of bisection for the search of solutions makes the approach rather inefficient, although various bisection rules are discussed. [Canny, 1984] transforms polytopes (that are translated along a straight line) into configuration space and represents rotations by quaternions thus staying purely algebraic. For two objects in motion he gets algebraic equalities and inequalities in a twelve dimensional configuration space. [Schwartz and Sharir, 1983] use cylindrical algebraic decomposition [Collins, 1975] to analyze the motion planning problem. They say that their method, "though hopelessly inefficient in practical terms, this result nevertheless serves to calibrate the computational complexity of the motion planning problem", i. e. it can be solved in polynomial time.

12.6 Example: Manufacturing Verification Using Linear Programming

For a suitable modeling, we approximate the objects in a work cell by polytopes and compute the convex hull thus getting *convex polytopes*. Furthermore, non-convex objects can be hierarchically structured and decomposed into convex elements. This form of approximation is sufficiently close to real objects since the lion's share of technical objects is composed of polytope-style objects. Moreover, the hierarchical modeling allows the fast detection of non-collision situations, in particular situations where even the convex hulls of the compound objects do not intersect.

As can be seen from Fig. 12.7b, the problem of finding a collision between two convex polytopes (2D polygons in the figure) is basically the same problem as the linear programming (LP) problem (Fig. 12.7a), see, e. g., [Cameron, 1985] for details. Thus, one can attack the collision problem using LP techniques.

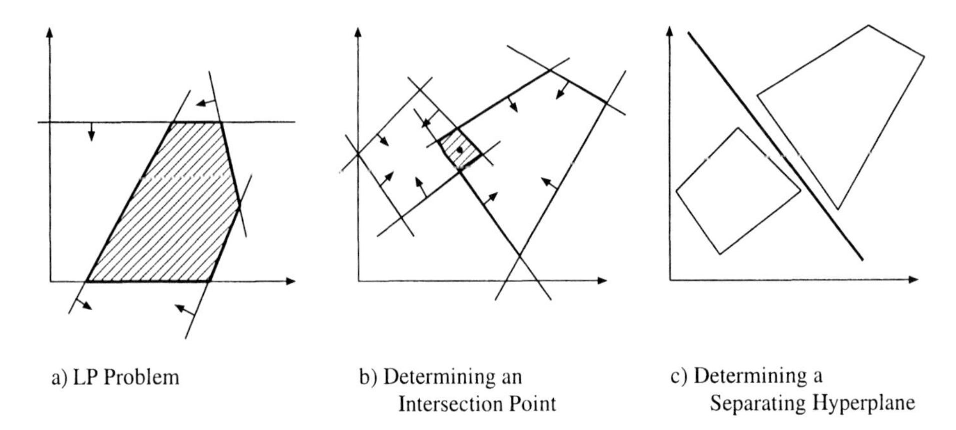

Figure 12.7: Formulating collision checking as a linear programming problem

Basically, we found two alternatives for translating the collision checking problems into an LP problem, namely:

- Using LP for determining a point of intersection, see Fig. 12.7b, and
- Using LP for determining a separating hyperplane, see Fig. 12.7c.

12.6.1 Formulating Collision Checking as Linear Programming Problems

12.6.1.1 Determining a Point of Intersection

One possibility to check, whether two objects do collide is to determine a point that is contained in both of the objects (a *point of intersection*). Thus, the corresponding version of the collision checking problem can be formulated in the following way (Fig. 12.8):

Problem P-IP "Determining a Point of Intersection":

Given: Two convex 3D polytopes $P_1 \subseteq \mathbf{R}^3$, $P_2 \subseteq \mathbf{R}^3$.

Find: $\vec{x}$, such that $\vec{x} \in P_1$ and $\vec{x} \in P_2$ (a point contained in both polytopes). If no such point exists, report this fact.

In order to formulate the problem of finding a point of intersection as an LP problem, we need the following

Theorem on the Intersection of Convex Sets:

Let P_1 and P_2 be subsets of $\mathbf{R}^n$. If both P_1 and P_2 are convex, $P_1 \cap P_2$ is also convex.

For a proof, see, e. g., [Cameron, 1985].

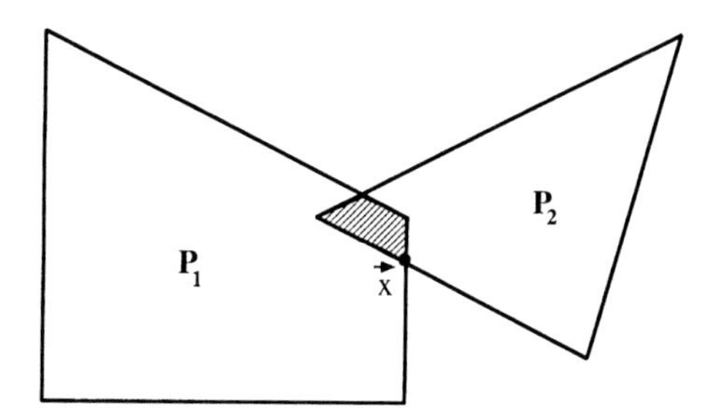

Figure 12.8: Determining a point of intersection

The intersection of the two polytopes is bounded, because both polytopes are bounded by definition. Thus, the intersection of two polytopes in $\mathbf{R}^n$ forms a polytope in $\mathbf{R}^n$.

Since every polytope is also a polyhedron, the facets of a polytope define the bounding hyperplanes of the half-spaces that describe the polytope by their intersection. Consequently, the two sets of half-spaces can be joined into a set of half-spaces defining the intersection of the two polytopes.

If this set of half-spaces is now considered as the restrictions for the feasible region of an LP problem, a point of intersection can be found by determining a point within the feasible region, and if no such region exists, by reporting this fact.

Thus, the corresponding LP version of the collision checking problem can be formulated in the following way (Fig. 12.9):

Problem P-IP-LP "Determining a Point of Intersection Using LP":

Given: $P_1 \subseteq \mathbf{R}^3$, specified by $\bigcap_{1\leq i\leq k_1} \langle \vec{a}_{1_i}, b_{1_i}\rangle_{\leq}$,
$P_2 \subseteq \mathbf{R}^3$, specified by $\bigcap_{1\leq i\leq k_2} \langle \vec{a}_{2_i}, b_{2_i}\rangle_{\leq}$
(two polytopes specified by their defining half-spaces).

Find: $\vec{x}$, such that $(\forall 1 \leq i \leq 2)((\forall 1 \leq j \leq k_i)(\vec{a}_{i_j}\vec{x} \leq b_{i_j}))$
(a point in the feasible region).
If no such point exists, report this fact.

When formulating the LP problem with the parameters of the restricting half-spaces as variables, the problem occurs that the values of the parameters may be positive or negative, whereas LP variables may only be positive in the standard formulation. This problem can be overcome by a common trick in linear programming: The input variable is split into its positive part and its negative part resulting in two variables in the LP formulation.

Now this adapted set – containing only variables that are non-negative – has to be simply input into an LP algorithm using $\vec{0}$ as the objective function to determine a point within the feasible region, if such a region exists. In our example, a possible answer is $(5, 1)$, witnessing that the two polytopes are intersecting. If the LP algorithm cannot determine such a point (reporting "no solution"), the two polytopes cannot intersect, i.e. the two corresponding objects cannot collide at the moment considered.

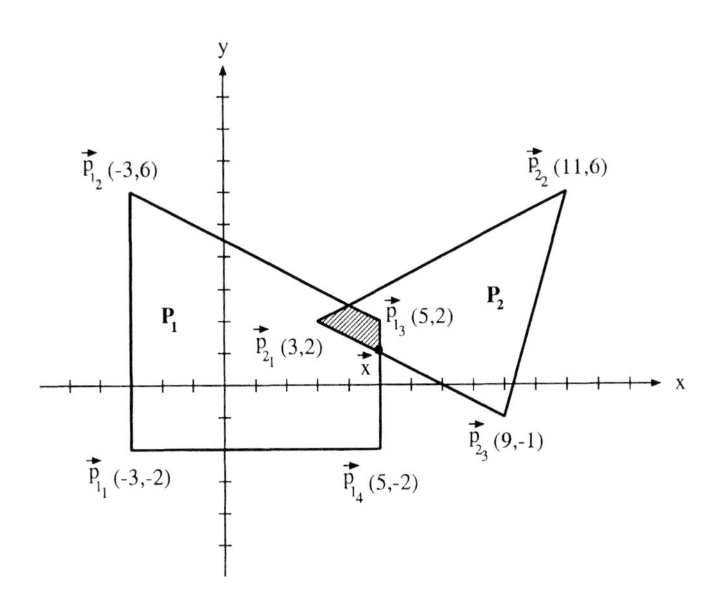

Figure 12.9: Determining a point of intersection using linear programming

For static collision checking only, a similar approach is pursued in [Duelen *et al.*, 1990]. However, they use only very simple enclosing polytopes and do not report on any improvement of this LP formulation, nor do they apply their approach to dynamic collision checking. In [Mayr, 1991b], we show that a closer look onto the problem results in a much better LP model and a corresponding speed-up in computation time.

12.6.1.2 Determining a Separating Hyperplane

We have already sketched the idea to perform collision checking by means of a hyperplane that separates the two objects to be checked. Such a hyperplane constitutes a plane certifying the disjointness of these objects. In this section we show how to generate a separating hyperplane, if one exists. We again represent the two objects by their enclosing convex polytopes. Thus, the problem we solve is the following:

Problem P-SH "Determining a Separating Hyperplane":

Given: Two convex 3D polytopes $P_1 \subseteq \mathbf{R}^3$, $P_2 \subseteq \mathbf{R}^3$.

Find: A hyperplane, specified by $\langle \vec{a}, b \rangle_=$, such that
$(\forall \vec{p}_1 \in P_1)(\vec{a}\vec{p}_1 < b)$ and $(\forall \vec{p}_2 \in P_2)(\vec{a}\vec{p}_2 > b)$.[3]
If no such plane exists, report this fact.

---

[3]It is not necessary to admit the alternative case swapping "<" and ">". This case can be achieved by multiplying $\vec{a}$ and b with -1.

If two convex polytopes do not intersect each other, there does exist a hyperplane that separates the two polytopes, i. e. one polytope is a subset of a half-space defined by the hyperplane, whereas the other polytope is a subset of the complement of this half-space.

The following strategy allows the formulation of the problem of determining a separating hyperplane as an LP problem: The hyperplane is described by its parameters $\vec{a}$ and b. These parameters are implicitly determined by the conditions that one polytope has to be on one side of the plane (i. e. contained in one half-space determined by the hyperplane), whereas the other polytope has to be on the other side of the plane (i. e. contained in the complement of the above half-space).

The values of the variables $\vec{a}$ and b may be positive or negative, so again we have to split up the parameters into their positive part and their negative part (as for Problem P-IP-LP). Then the problem of determining a separating hyperplane can be formulated as an LP problem in the following way (using the value ε to guarantee that the two polytopes are, at least, at a certain security distance ε from each other)[4]:

Problem P-SH-LP "Determining a Separating Hyperplane Using LP":

Given: $P_1 \subseteq \mathbf{R}^3$, specified by $convexHull(\vec{p}_{1_1}, \ldots, \vec{p}_{1_k})$,
$P_2 \subseteq \mathbf{R}^3$, specified by $convexHull(\vec{p}_{2_1}, \ldots, \vec{p}_{2_l})$,
$\varepsilon > 0$ (a security distance).

Find: $\vec{n} := \left[n^+_{\vec{a}_x}, n^+_{\vec{a}_y}, n^+_{\vec{a}_z}, n^+_b, n^-_{\vec{a}_x}, n^-_{\vec{a}_y}, n^-_{\vec{a}_z}, n^-_b\right]$, such that
$n^+_i \geq 0,\ n^-_i \geq 0,$

$$\vec{a} := \begin{bmatrix} n^+_{\vec{a}_x} & - & n^-_{\vec{a}_x} \\ n^+_{\vec{a}_y} & - & n^-_{\vec{a}_y} \\ n^+_{\vec{a}_z} & - & n^-_{\vec{a}_z} \end{bmatrix},\ b := n^+_b - n^-_b,\ and$$

$(\forall 1 \leq i \leq k)(\vec{a}\vec{p}_{1_i} \leq b - \varepsilon),\ (\forall 1 \leq j \leq l)(\vec{a}\vec{p}_{2_j} \geq b + \varepsilon).$

If no such $\vec{n}$ exists, report this fact.

This problem can be translated into an LP matrix description in the following way:

$$\begin{bmatrix} -\vec{p}_{1_1,x} & -\vec{p}_{1_1,y} & -\vec{p}_{1_1,z} & 1 & \vec{p}_{1_1,x} & \vec{p}_{1_1,y} & \vec{p}_{1_1,z} & -1 \\ \vdots & & & & & & & \vdots \\ -\vec{p}_{1_k,x} & -\vec{p}_{1_k,y} & -\vec{p}_{1_k,z} & 1 & \vec{p}_{1_k,x} & \vec{p}_{1_k,y} & \vec{p}_{1_k,z} & -1 \\ \vec{p}_{2_1,x} & \vec{p}_{2_1,y} & \vec{p}_{2_1,z} & -1 & -\vec{p}_{2_1,x} & -\vec{p}_{2_1,y} & -\vec{p}_{2_1,z} & 1 \\ \vdots & & & & & & & \vdots \\ \vec{p}_{2_l,x} & \vec{p}_{2_l,y} & \vec{p}_{2_l,z} & -1 & -\vec{p}_{2_l,x} & -\vec{p}_{2_l,y} & -\vec{p}_{2_l,z} & 1 \end{bmatrix} \vec{n} \geq \begin{bmatrix} \varepsilon \\ \vdots \\ \varepsilon \\ \varepsilon \\ \vdots \\ \varepsilon \end{bmatrix},$$

$$\vec{0}\,\vec{n} \longrightarrow MIN.$$

[4]For achieving a safe task, a rather big ε should be chosen, in this way avoiding critical situations due to non-considered factors, e. g., dynamics. However, in order to increase the chance of finding a separating hyperplane, ε should be rather small. In practical applications, a few millimeters turned out to be a good choice.

To give an example, we change the scenario of Fig. 12.9 by translating polytope P_2 according to the vector $(0, 3)$. In doing so, the two polytopes do no longer intersect, which allows the determination of a separating hyperplane (Fig. 12.10).

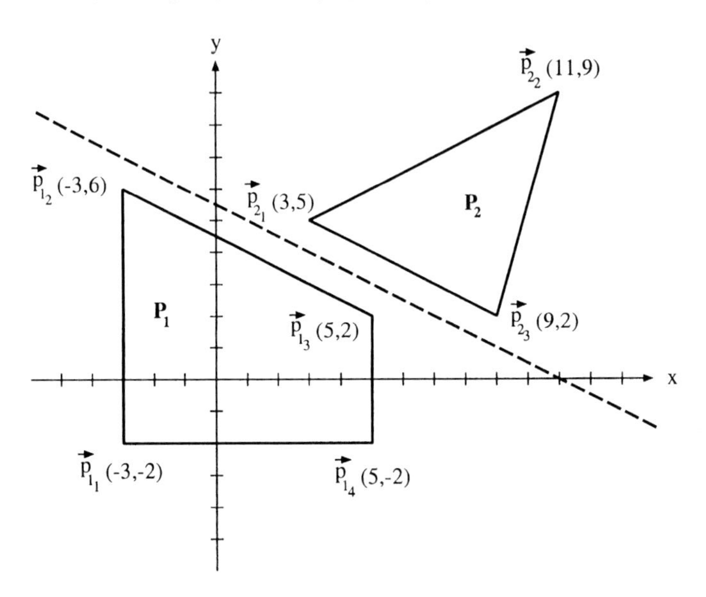

Figure 12.10: Determining a separating hyperplane using linear programming

Compared with the determination of an intersection point, the determination of a separating hyperplane is more time-intensive due to the larger LP matrix. However, if a separating hyperplane can be found, a forecast can be made, how long no further static collision checks are necessary while still guaranteeing collision-freeness of the whole task.

12.6.2 Different Representations of Polyhedra

In our formulations of the last section, both polytopes have been specified by a set of vertices. However, as already mentioned, polytopes may alternatively be specified by their restricting half-spaces. Furthermore, unbounded polyhedra can only be specified by their restricting half-spaces. Such situations may also occur in algorithmic verification of a manufacturing process, e. g., when modeling a joint restriction of a robot that allows only motions in the 3D half-space with positive x-coordinates.

We refer the reader to [Mayr, 1991b], where we show how collision checking of polyhedra can be modeled as LP problems that are represented using half-spaces. Also, the LP modeling of a mixed representation (one polyhedron specified by its vertices, the other specified using half-spaces) is discussed there.

12.6.3 Tailoring the Linear Programming Formulations

LP algorithms are generally rather insensitive to expanding the problem by adding a few columns or rows. To solve the dynamic intersection problem, we can therefore proceed as follows: We approximate the path of the objects in a piecemeal linear way with respect to velocity. Now we can derive time intervals where both objects move linearly by the time and can be described using a single transformation matrix. If we now add the time as a fourth dimension to our 3-space and sweep our objects along the moving line in the time axis, we get 3D objects extruded into 4D prisms. If the 3D objects are convex it is obviously that the 4D prisms are convex either.

For this reason we can perform the same LP transformation as in the static case, only taking the 4D vertices for the LP formulation.

By weighing the vertices with $t_0 = 0$ as more important than those with $t_0 = t$ (assuming that we test for the interval $[0 \dots t]$) we can utilize the minimization facility of the LP algorithm to determine the lowest (first) t_0 where a collision can be detected. Doing so, we can get the first moment with collision together with the collision point with a single collision check.

All our LP checks solve the static collision checking problem (using 3D polytopes) as well as the dynamic collision checking problem (using 4D polytopes) completely. Thus, for the different problems one can choose the alternative that suits best (i. e. that leads to the shortest computation time).

12.6.4 Algorithmic Analyses and Improvements

The different formulations of the collision checking problem as an LP problem result in LP matrices of different size. Experiments have shown that in our applications the sizes of the LP matrices correlate with the computation times in a linear way. In Table 12.1, we therefore list the size of the LP matrices of the different problem formulations ("original formulation"). We present the static (3D) case only, because the ratio between the different formulations remains the same in the dynamic case; only the one extra dimension and the larger number of vertices has to be considered.

The size of the input is characterized by the total number n of vertices (or facets, depending on the description) of the two polyhedra to be checked.

The following facts of our LP matrices can be used for improving our formulations:

- Most LP matrices representing our collision checking problems are far from being square matrices. They have a rectangular shape that gets the more rectangular the more elements are used to describe the input. This special structure can be utilized for an increase of efficiency.
- A $\vec{0}$ objective function can be eliminated using one of the constraints as objective function and testing the feasibility of the optimal value with respect to this new objective function.
- When determining a separating hyperplane instead of generating an arbitrary hyperplane, the homogeneous form of this hyperplane (i. e., $b = 1$) can be determined, thus eliminating the matrix columns representing the variable b.
- In certain formulations, the variables that are necessary to get a standard LP problem (auxiliary variables) considerably contribute to the total size of the LP matrix.

Consequently, one should try to re-formulate the problems, such that the number of internally used variables is decreased, e. g. by dualizing the problems.

These ideas lead to a reduction of the sizes of the LP matrices as shown in column "improved formulation" of Table 12.1.

Problem	Original Formulation	Strategies	Improved Formulation
P-IP-LP	$4n^2 + 12n$	Dualization	$12n + 36$
P-IP-LP$_{BC}$	$10n + 25$	-	$10n + 25$
P-SH-LP	$8n^2 + 16n$	Hom. HP, Dual.	$12n + 36$
P-IP-LP$_{HS}$	$4n^2 + 12n$	Dualization	$12n + 36$
P-SH-LP$_{HS}$	$8n + 16$	Elim. $\vec{0}$ Obj. F.	$6n + 9$
P-IP-LP$_{M}$	$2n^2 + 3n + 1$	Elim. $\vec{0}$ Obj. F.	$2n^2 - 3n + 1$
P-SH-LP$_{M}$	$3n^2$	Elim. $\vec{0}$ Obj. F.	$3n^2 - 5n + 2$

Table 12.1: Matrix sizes of different LP formulations of collision checking

The sizes of the LP matrices can be used to choose the most efficient LP formulation for the different collision checking alternatives. One, however, has to keep in mind that some of the improvements can only be applied under certain conditions (e. g., the objective function has to be $\vec{0}$ in order to be able to eliminate it). Also the level of information that can be gained about a collision situation may be different for the different formulations.

We have implemented the presented collision checking procedures in the frame of a simulation system in standard C language in such a way that it runs on standard PCs or workstations. Generally one can say that our algorithms decide the collision problem very fast [Mayr, 1990], which allows real time collision checking during the whole graphic simulation process. The extension to algorithmic task verification by dynamic checking is possible in a very elegant and simple way. Furthermore, the step into dynamic checking does not lead to disastrous calculation times. Here, a comparison with other collision checks definitely proves our approach superior.

Chapter 13

Virtual Environments for Automation

This chapter informs on the transition from conventional simulation systems to (distributed) virtual automation environments. Suitable system architectures and information flow models (e. g. software sensors) are discussed, and the concept of a virtual factory is introduced. Examples from various different areas illustrate the breadth of the impact of virtual automation environments.

13.1 From Simulation Systems to Virtual Worlds

13.1.1 Distributed Virtual Reality

Virtual reality (VR) and cyberspace are vogue words in big business today. Almost every computer game is a "cybergame" and deals with distributed virtual worlds. A new vogue word is born: *distributed virtual reality*. [Roehl, 1995] defines distributed virtual reality as follows: "The idea behind distributed VR is very simple; a simulated world runs not on one computer system, but on several. The computers are connected over a network (possibly the global Internet) and people using those computers are able to interact in real time, sharing the same world."

In most applications distributed VR follows a naive approach of a simple client / server model. The server is responsible for the whole world and only sends update packages to the clients. The clients have only a copy of the world and change the world each time a package reaches the client. However, distributed VR is much more: "Distributed" is a hint that the virtual world is distributed on several machines physically. For example, in architectural applications a house is divided into several rooms. Normally the whole house is stored physically on one computer. The new approach in distributed VR is to divide the house and store the rooms on different computers. The consequence of this approach is to keep a consistent world and to find an effective algorithm that handles all the messages and traffic on the net. The crucial point of virtual reality is the interaction in real time.

13.1.2 Client / Server versus Peer-to-Peer Architectures

A modern, distributed simulation environment for virtual reality contains two main challenges for a modeling system, supplying the world objects with knowledge about the environment in order to enable "intelligent" behavior, and disentangling the communication in order to decentralize the responses to environmental events.

The first implementations of simulation systems and also virtual reality environments have been done on a single-user / single-system basis. However, with the increasing need of adapting a virtual world during a simulation run and the rising interest in virtual cooperation of two or more users, concepts for distributed use of VR systems had to be developed.

Due to the popularity of client-server techniques in the beginning of the Nineties, a big share of the current distributed virtual reality implementations use a *client-server architecture* [Greenhalgh, 1994]. In this architecture, a server computer runs the simulation of a virtual world. Participants can interact with each other and with this world via client computers that are connected to the server by wide-area networks. Each client maintains a small cached subset of the server database that is displayed by the client's rendering engine. In this way, actually not the simulation itself, but the access to it is distributed. While this approach considerably simplifies system design, since there is only one copy of the simulated world, it does not scale for global use. For many clients the central server soon becomes a bottleneck.

Using a *peer-to-peer architecture*, the simulation of the world is not performed by a single computer system, but is distributed among several hosts connected over wide-area networks. Again people using these computers are able to interact, sharing the same virtual world. Only this approach provides the potential for global scalability. However it raises the problem of maintaining consistency of the simulation under the restrictions of limited network bandwidth.

Therefore a combined peer-to-peer network of client-server clusters is proposed as the suitable architecture, since for an applicable virtual reality system, two crucial requirements have to be fulfilled:

1. The *global consistency* of the simulation must be guaranteed.
2. The system must show *real-time behavior* in order to get a feeling of reality (demand for "quality of service", see [Nahrstedt and Steinmetz, 1995]).

In order to realize the virtual simulation scenario a network of simulation hosts ("servers") implements a shared space of objects with autonomous behavior. Each server handles a set of "areas of interest" corresponding to the virtual locations of those participants that are connected to this server. All servers may execute object behaviors concurrently and independently. Object copies are generated and discarded according to the servers' varying areas of interest as participants move through the virtual world.

13.2 Sensor-based Simulation

13.2.1 Simulation With Local Intelligence

In a distributed world model like that discussed in the previous chapter, the representation of the scenario is done by components that are held distributedly (and made persistent using appropriate databases). The key issue is that it is not enough to just keep the components distributed and handle the communication centralized. Depending on the different goals, levels and areas of communication, also the information transfer (e. g. event handling) has to be distributed.

The author, together with students, has developed an object oriented framework for such an intelligent simulation environment that focuses on motion controlled by virtual software sensors enabling intelligent local object behavior by means of object-oriented techniques for message handling, see [Peneder, 1996], and [Wolf, 1996]. Local, distributed managers for information handling and reaction (*sensor managers*) allow quick responses and environmentally correct behavior of objects, thus reducing the amount of information that has to be processed in a central manager (i. e. main event handler) to a manageable level. Only these concepts allow a simulation, e. g., in a computer net without creating a crucial bottleneck of one single information and communication handling process.

In order to decrease (communication) complexity, objects must have an internal behavior and communicate by exchanging messages ("behavior-integrated objects", see, e. g., [Kowal, 1992]). This paradigm has meanwhile become well studied in the frame of object oriented programming. Objects thus become intelligent "autonomous agents" that act in the virtual world and may have their own impact onto the virtual world (in the same way as introduced in the theory of autonomous systems, see, e. g., [Agha, 1986]).

Both ideas, decentralizing communication as well as subdividing scenarios into areas of interest are already well studied in the literature, see, e. g., [Macedonia *et al.*, 1995]. However, till now these ideas have been incorporated only rarely – if at all – into manufacturing planning and simulation systems actually used at the shop floor. Our new approach is to use these concepts – in a somewhat scaled-down way – in order to enhance classical manufacturing simulation systems such that they can cope with modern, far more complex manufacturing processes and take use of up-to-date multi-processing computer networks. This approach allows to migrate from standard manufacturing simulation to *virtual automation environments*, beginning with the modeling data available and enhancing component by component.

13.2.2 Creating Intelligent Objects by Adding Sensor Information

In the case of a virtual factory autonomous agents represent robots or work cells that perform certain activities like production or transportation of goods. However, it is not sufficient to let objects act just in the way their task is planned in advance. Spontaneous, unplanned events (like an interruption by a human "intruder" in the scenario) cannot be coped with in that way. Therefore, objects must be able to get information about their relevant environment in the moment they need it.

This can be achieved by modeling suitable *software sensors* for each of the interacting objects.

This step into a second generation of mobility [Desoyer *et al.*, 1985] has been proposed (and implemented) for hardware devices already for several years. Modeling based upon sensors has already been studied in behavioral simulation [Wilhelms and Skinner, 1990] and has been sketched for creating generic virtual reality systems [Torguet and Caubet, 1995]. However, task planning in off-line simulation systems even nowadays mostly lacks the use of sensors for getting "live" runtime information about an object's virtual environment.

Consequently, the main step in improving a simulation system is the optional addition of software sensors to each of the objects in the scenario in the way shown in Fig. 13.1, achieving "intelligent" objects.

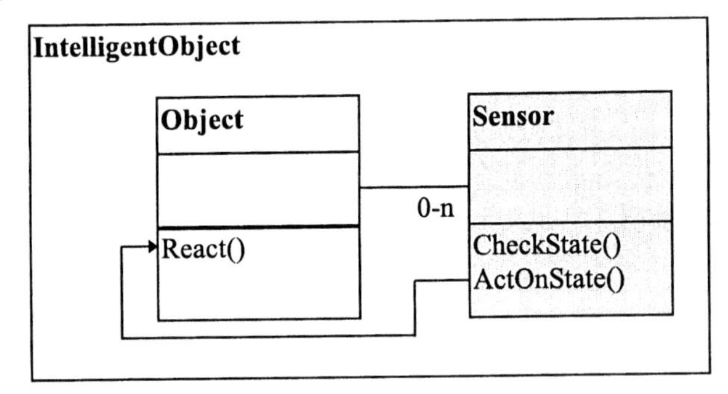

Figure 13.1: Creating intelligent objects by adding sensors

An object is assigned a number of sensors in order to be able to explore the environment and react properly. Each sensor can check the state of its corresponding environment objects (`CheckState`) and knows the proper action for each state (`ActOnState`). This action leads to the corresponding reaction (`React`) of the object.

It is essential to be able to extend any ordinary object to an intelligent object by adding sensors and behavior, but this extension should be optional and even be possible during the design of a scenario. Otherwise, scenarios either immediately become too complex with respect to their communication structure, or they are too inflexible and restricted. We found that the object-oriented paradigm supplies excellent concepts for such an optional extension and strongly recommend a sound object oriented implementation of a world modeler including sensor models.

13.2.3 Creating Intelligent Scenarios by Adding Sensor Managers

The introduction of sensors into a world model allows "intelligent" autonomous actions of objects depending on their current circumstances. However, each of the sensors basically communicates by requesting information from other objects and sending messages to its owner object. This leads to a considerable increase in communication efforts, i. e. the number of events that have to be handled in event-driven systems.

If one single event handler (the "main" event handler) has to handle all these events, a new bottleneck in the virtual reality system is being created that might be tougher than the ones eliminated. As stated in, e. g., [Ghee, 1995], most actual prototypes of distributed virtual reality systems use an object-oriented world description model, but keep the message communication and the simulation mechanism external, i. e. global. Objects have some form of local behavior, but the simulation is actually controlled by external processes.

In order to overcome this communication bottleneck, we suggest to split the world into an appropriate number of scenarios. Such scenarios can be either geographically separated ("world zones", see [Roehl, 1995]) or be defined by a different layer of information quality ("areas of interest"). If one considers the model of a virtual city, geographically separated scenarios could be, e. g., the single streets, whereas scenarios defined by areas of interest could be classes of traffic or types of tourist routes.

For each of such scenarios, the intra-scenario communication can be handled at an intermediate layer between (object-)local information and (world-) global event communication. In the field of software sensors a *sensor manager* can be implemented for each of the scenarios (subsections of the world model), creating "intelligent" scenarios, see Fig. 13.2. Only inter-scenario communication has to be handled on world level.

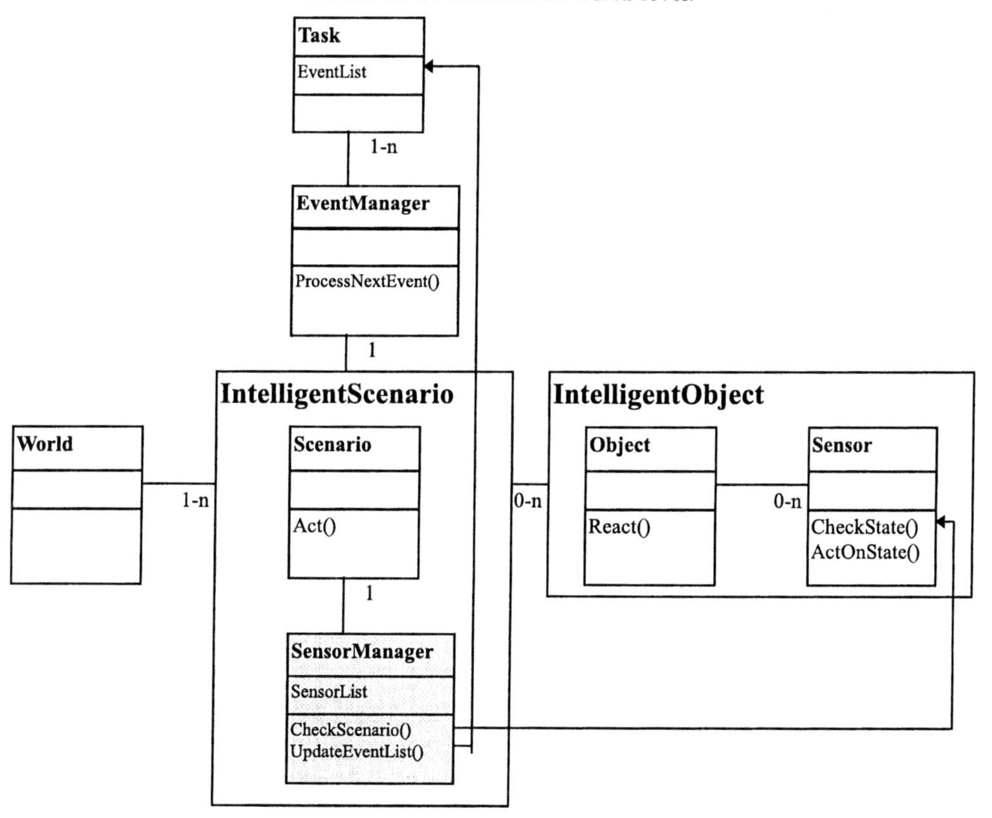

Figure 13.2: Creating intelligent scenarios by adding sensor managers

Such sensor managers reduce the amount of information that has to be distributed throughout the whole simulation by coordinating the checking of states and arrange for all actions that can be performed in the frame of each single considered scenario.

A similar conclusion must be drawn for sensor managers as that for sensors: It is essential for the applicability of the concept that any scenario can be optionally equipped with a sensor manager. We assume that several other communication layers like that of sensors for controlling motions can be reduced with respect to complexity by introducing managers in a similar way like sensor managers.

13.2.4 Algorithmic Modeling of Sensors

The implementation of software sensors presented in this chapter origins from the concepts for real-time collision checking developed by the author, based upon linear programming as described in Ch. 12.6 (see also [Mayr, 1991b]). The original focus of this application was the modeling of sensors for off-line manufacturing simulation and programming in order to enable off-line teaching and path following. By extending the applicability of the simulation system and adding interactive and cooperative (distributed) features, also the types of modeled sensors have been extended. Since our algorithms also determine some quality of non-intersection (minimal distance), we can additionally model the class of approach sensors for, e. g., object configuration and autonomous motion. By utilizing the optimization functionality of the underlying LP method, the prediction of times and locations of (future) collisions is also possible. Still, the main functionality of the sensors implemented in our simulation system is the improvement of motion security and motion quality.

For our first applications of software sensors, the graphic simulation system developed by the author [Mayr, 1991b] was extended in order to allow uncertainty handling (like compensating manufacturing tolerances) and defect handling (like detecting broken tools), bringing simulation closer to modeling the real world instead of modeling idealized scenarios. After several re-engineering phases of the class libraries, the simulation framework gained this way is being successfully used as a basis for projects in other areas concerned with motion modeling and simulation. For a detailed description of the core model library of our system, we refer to [Peneder, 1996].

The following classes of sensors are available in our sensor library. The classification is according to [Desoyer *et al.*, 1985].

Approach Sensors

Distance sensors measure translational distances or rotational distances between two objects or reference frames. They are subdivided into *range sensors*, measuring the minimal distance between two objects, and *angle sensors*, measuring the angle between the reference planes of two objects. Range sensors can be realized utilizing the distance measuring capability in the way described above. Angle sensors can be realized measuring distances with respect to suitable frames and applying trigonometric conversions. Examples are echolot, radar, laser tracking system.

Presence sensors determine, whether a defined subspace contains any object. They are realized by modeling the (invisible) subspace of the checking range of the sensor in an adequate way (e. g., the light beam of a light barrier) and performing a collision check with each possible candidate for object presence. Examples are light barrier, magnetic sensor.

Position Sensors

Absolute position sensors determine the position of an object in the scenario. *Relative position sensors* determine the position of an object relative to a second object.

Position sensors are realized by taking the whole considered object as a feasible solution and "optimizing" in order to get the current location of the reference frame, either with respect to the origin of the model world or with respect to the origin of the second object. The origin of the second object (in absolute values) can be determined in the same way.

State Sensors

Discrete state sensors determine the state of an object delivering discrete information. Examples of such objects are traffic light, control switch. *Continuous state sensors* determine the state of an object delivering continuous information. Examples of such objects are light dimmers, car speed pedal.

The state of a discrete state sensor can be determined directly by reading the current state from the sensor. Of course, the same principle could be applied to continuous state sensors (i. e. reading the value of the status of the sensor). Frequently, however, this status, although correct in its relative value, is not transferable into absolute values without checking for environment parameters. An example would be the tool feed in an NC machine which is specified by the motion of the tool mounting position. During machining, however, the tool length shrinks due to usage, and thus the feed has to be corrected appropriately in order to guarantee correct tool impact. Distance determination is frequently used for such corrections, called *calibrating* of continuous state sensors.

Tactile Sensors

Contact sensors determine, whether two objects are in contact with each other or intersect each other. *Location sensors* additionally determine the (one) location of the contact or intersection.

Tactile sensors constitute the original application of collision checking, namely detecting contacts and intersections. Appropriate optimization is used for realizing the location determination of location sensors.

13.3 Virtual Factories

13.3.1 The Concept of a Virtual Factory

When the term *virtual factory* is used in the literature, it comprises different backgrounds and meanings. In literature the spectrum of virtual factories covers enterprises that employ only tele workers as well as enterprises that offer a "virtual product", see [Schräder, 1996]. The very loose and informal alliance of enterprises is one comprehension: "The virtual factory corporation is a temporary network of independent companies - suppliers, customers, even erstwhile rivals - linked by information technology to share skills, costs, and access to one another's markets. It will have neither central office nor organization chart. It will have no hierarchy, no vertical integration. ... In the concept's purest form, each company that links with others to create a virtual corporation will be stripped to its essence. It will contribute only what it regards as its core competencies." [Byrne *et al.*, 1993]

The main benefits of this evolutionary way of manufacturing expanding towards virtuality are:

- Partners can be incorporated at any stage of a relationship.
- Partners can be incorporated with varying degrees of technology sophistication.
- All required functionality is provided to support every application across the network.

A different aspect is put to foreground in [Smith, 1996], focusing on virtuality by process simulation: "Virtual factory allows the user to quickly prototype complex processes, optimize product characteristics, provide schedules and cost analysis, and review all project documentation, online."

Regardless of the exact specification, the common divisor of all the definitions is gaining flexibility by means of modern information technology, see [Davidow and Malone, 1992], frequently focusing on utilizing this flexibility – and its lack of risk – for teaching purposes.

13.3.2 Need for Virtual Factories

The aim of virtual factories is to depict as many processes as possible of a real factory with the purpose to create an interactive training and education environment, based on the newest marketing strategies, software developing techniques, and business management knowledge. For this reason, the whole system is split into modules that communicate via (inter)net facilities and in this way enable an online distributed learning facility, resulting in a complete production simulation, its control, and also the visualization of the simulated production events on several workstations.

Besides the advantage of training at no financial risk of damages, supplied by the use of software simulation, the key advantage for virtual factories is the use of distributed simulation. Many companies have their production spread over several locations, with just one training center at a central location. Although there are several of the production cells available at the training center, at any time at least some of the production cells have to be simulated by virtual (software) production cells and buffers for the real parts. The goal is to simulate the whole factory at any subset of the locations, using real-world machines or virtual-world simulated machines as need arises for simulating special situations or due to the availability of certain cells only at a distant location.

13.3.3 Using Virtual Factories for Education and Training

A virtual education environment enables the creation and the simulation of all major departments of various production companies. The trainers can act as customers or sub-contractors, while the students plan, coordinate, execute and check all the tasks necessary to run the company. In this way the students learn to react on changes, solve problems and master crises within the company i. e. the team. When finishing the simulation, results can be measured and analyzed, like the profit gained, or the quality achieved. With this approach, terms like team work, information flow, etc. shall become meaningful to the students in the frame of a practical context. Additionally, students shall be motivated to investigate the complex structure of the information flow within a real company.

A virtual factory can be used to train teachers, company managers, and workers during short courses in the frame of lifelong learning. The main use, however, is to educate students in understanding the whole function of the factory in different kinds of institutes. It can also be used in to train the skills of understanding the whole entity of functions of the factory between order and delivery. Such systems are generally open and distant learning environments which also enable the substitution of parts of the production process of a company. This allows the training of employees in distributed companies at one location, utilizing the facilities available at that location and simulating the facilities available in reality only at other locations [Mayr, 1997b].

13.4 Example: Virtual Toy Train

13.4.1 The Virtual Toy Train Scenario

In order to teach the problems arising in distributed simulations, a scenario consisting of virtual toy trains has been developed by some of the author's students (see [Marko, 1996], [Peneder, 1996], [Wolf, 1996]). In the scenario, two trains follow the same track with individual speeds. A signal system can be used for stopping trains at a railway station. The system can be used in two ways:

1. Each train is controlled by a different human user. The signal system may be controlled by one of these users or by a third one.
2. One of the trains is controlled by the user; the other one is controlled by the computer. The signal system may be controlled by the user controlling the train or by another user.

13.4.2 Modeling a Toy Train Using Software Sensors

Both trains have a (discrete state) sensor that checks for the signal system and a second (range) sensor that prevents one train from running into the other. The range sensor is a good example for the frequently occurring principle that the reaction on a positive sensor result (i. e. clash of the trains) depends on the behavior of both objects involved (i. e., depending on the state of one train, the other either has to reduce its velocity or get to a complete stop). The sensor model for the single-user mode is depicted in Fig. 13.3.

The virtual toy train is used for two purposes:

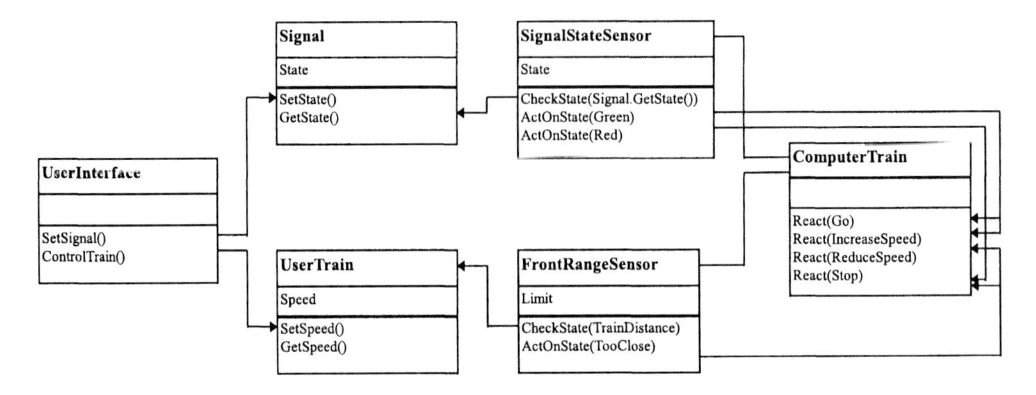

Figure 13.3: Sensor model of the toy train scenario "user vs. computer"

1. Teaching students how to integrate objects into a simulation scenario, in particular intelligent objects that include sensors. An example for a current extension is a track following system that will allow a train to follow an arbitrarily designed track composed interactively during runtime.

2. Teaching students how to distribute control of a simulation system in a computer network. Currently, for this example a client-server solution is being used; a peer-to-peer expansion for larger clusters of scenarios is currently being designed in accordance with our project on virtual factory education, discussed in Ch. 13.6.

A sample constellation of the virtual toy train can be seen in Fig. 13.4.

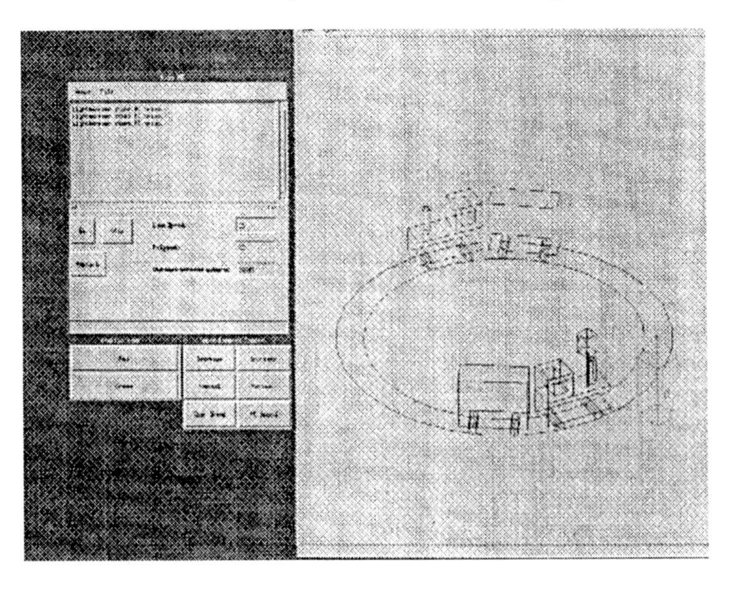

Figure 13.4: A sample virtual toy train scenario

13.5 Example: Turtle Neck Analysis

This application differs from the conventional approach of kinematic modeling by focusing on modeling biological aspects. The goal is to analyze the kinematic description of a long-necked turtle species, pleurodira (see Fig. 13.5). These turtles do not retract their neck straight, they bend it to the left or to the right side. This kind of turtles have a very flexible neck which allows the turtle to move the neck very fast and to strike the haul with tremendous speed.

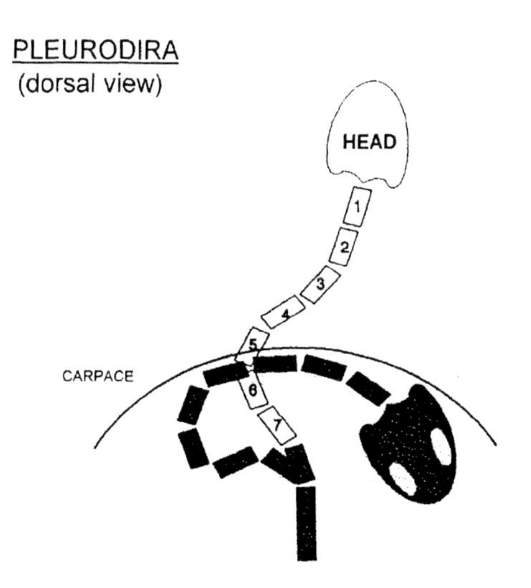

Figure 13.5: A systematic representation of a side necked turtle (pleurodira)

The flexibility of the neck is all the more amazing since the length of the neck is half the length of the turtle. At the Universitaire Instelling Antwerpen in Belgium these long-necked species of turtles are studied at the department of biology in order to gain insight for improving endoscopic medical devices by applying bionic methods [van Damme, 1993].

13.5.1 Kinematic Description of a Turtle Neck

In Fig. 13.5 all eight vertebrae are marked. In our application we restrict to a two-dimensional movement in the XY-plane, because the movement in the z-axis is not relevant for our studies [Peneder, 1996]. The A matrices for all links are the same and look as follows:

$$A_i = \begin{bmatrix} \cos\theta_i & -\sin\theta_i & 0 & 0 \\ \sin\theta_i & \cos\theta_i & 0 & 0 \\ 0 & 0 & 1 & d_i \\ 0 & 0 & 0 & 1 \end{bmatrix}$$

The Denavit-Hartenberg parameters are given in Tab. 13.1.

Link	a	α	d	θ
i	0	$0°$	d_i	θ_i

Table 13.1: Link parameters for the turtle neck

13.5.2 Modeling of a Turtle Neck

The implementation of the turtle neck follows the same way as described in the toy train application in the last chapter. The class `Scenario` possesses the class `Turtleneck`. `Turtleneck` is responsible for the kinematic model of the neck. The geometric representation of the neck is done by two cylinders, one in a vertical and the other in a horizontal position. All eight vertebrae have different ranges of rotation.

The goal of our investigation is to gain more insight into the interaction of the vertebrae in order to allow a quick and flexible extension and retraction of a turtle's head. We hope to be able to derive technical knowledge for improving the versatility and reducing the size of medical instruments for endoscopy and gastroscopy. The analysis of the relative position of the vertebrae with respect to each other is done using numerous appropriate sensors.

A picture of the motility analysis of the turtle neck can be seen in Fig. 13.6. For exact solutions of the problem and implementation details, we refer to [Brunner, 1996].

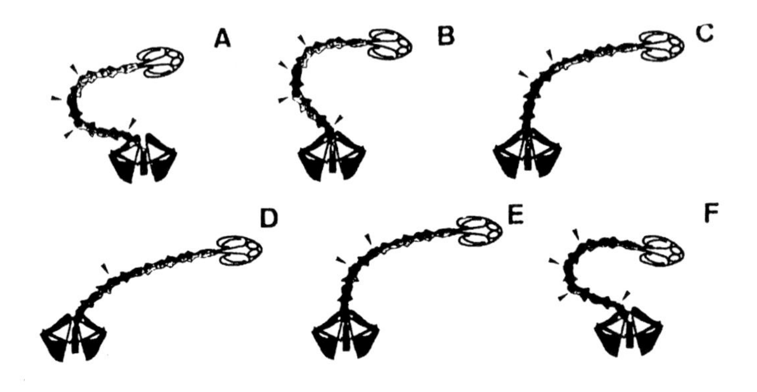

Figure 13.6: Motility analysis of the turtle neck

13.6 Example: The ViFE System

13.6.1 ViFE – A Virtual Factory for Distributed (Tele-)Education

The project *Virtual Factory for Education* (*ViFE*) was initiated by a Finnish consortium [Karppanen, 1996] together with partners from 6 European countries and was granted a three years' financial support by the European Commission starting Dec. 1996. The purpose of the project is to make available an education environment for industrial engineering and management education [Karppanen, 1996]. The idea is to depict as many processes as possible of a factory floor. These processes can be operated either manually (by the course participant) or by simulators. So, if there is only one person acting on the ViFE (for example in the production department), all other processes can be simulated, but it is also possible to hold courses with a multitude of students over the Internet.

All of the simulators can be customized and adapted by parameters. Thus, specific situations, e. g., machine failures or the loss of working capacity, can be probed. Therefore a course with the ViFE can be held with different levels of difficulty. As decided at the first ViFE meeting [Toppinen, 1997], the contents of the ViFE project are:

- developing a distant learning system based on ISDN and Internet,
- developing an open learning system including initial vocational learning,
- developing measure gauges of team working ability of the educated groups,
- developing simulators for production, sales, and markets,
- modeling the logistic of the enterprise,
- modeling the action chain management method,
- collecting the results on a CD-ROM for dissemination and transfer.

The goal of the ViFE project is to develop a virtual factory that includes a production control system (PCS), a production planning system (PPS), and different simulators (market simulator, production simulator, etc.) as well as databases of the factory. Simulators describe the main functions of the factory, whereby value is set on utilizing modern virtual reality concepts (see, e. g., [Loeffler and Anderson, 1994]). The PCS and the PPS control the basic activities according to the action chain philosophy starting from the order to the delivery. The PCS, the PPS, and the simulators make use of the common database, the development of which is an essential part of the project. The PCS and the PPS can make use of multimedia pages from different functions of the factory. Also a major task is to develop training units and plans and training material for institutes and to transfer the material internationally.

The learning environment is suitable for open learning where the instructors guide through the ViFE system for distant learning with the help of Internet. ISDN is used to connect two or more institutes together. This already has been tested several times by the coordinator in Finland.

At Pohjois-Savo Polytechnic, Varkaus, Finland, a first prototype for ViFE is installed for evaluation and preliminary training purposes. The virtual teaching environment is set up in two adjacent rooms, with a third room serving as a real production line, see Fig. 13.7. With the progress of ViFE, also the production line will be available in virtual form, thus making the whole education system independent of stationary components. This enables a distributed education using ViFE at any computer network linked to the Internet. It also allows the comparison of the performance of different student groups, which should raise the motivation of the students considerably. Additionally, results can be stored in a database for later evaluation and comparison in order to trace the degree of improvement of the students' skills.

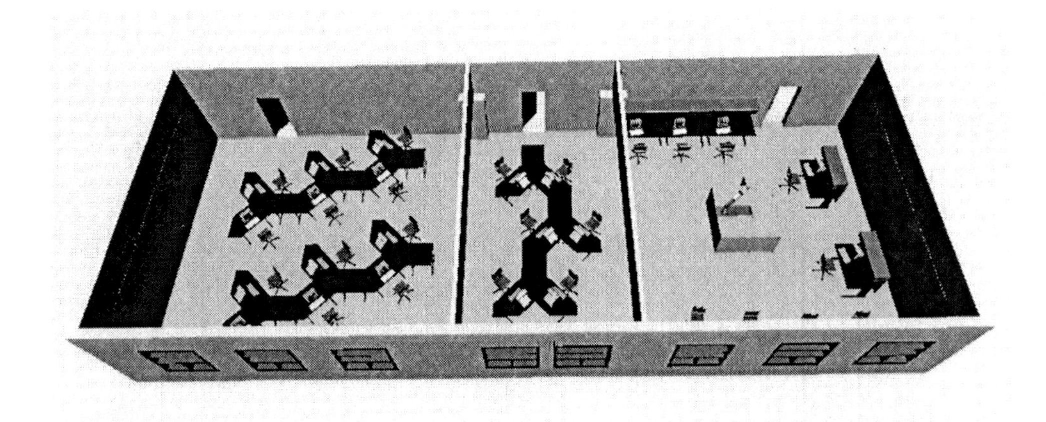

Figure 13.7: A "one location" ViFE setup

13.6.2 Existing Components

The VIFTOO System for Production Planning and Scheduling

As a preparation for the project, the Finnish coordinator supplied a rudimentary prototype for simulating a virtual factory, called VIFTOO (Finnish: Virtual Factory Toiminnan Ohjaus Ohjelmisto; i. e. virtual factory process control program), see [Surakka and Pulkrab, 1997]. VIFTOO is a client / server application, with a minimum type of server, namely the database. At least six PC's connected via a LAN are needed. If electronic mail is used for internal communication, each participant needs his own e-mail address. One PC (preferably the one of the course instructor) acts as system time server to adjust the other PC system times automatically.

In Finland, the system is installed in a large room that can be divided into areas ("divisions") for office automation (sales, purchase, production planning) and production operations (production, stock, manufacturing).

Education Using VIFTOO

The virtual factory is controlled by workstations that are connected via a LAN. The course instructors sit in a different room and communicate with the participants by mobile phone or e-mail. They take the position of the clients and suppliers. Depending on the number of participants one to three students work in a team. The participants are divided into at least six teams, being responsible for sales, production planning, purchase & stock, delivery & invoice, production, and manufacture. Each participant opens the mask corresponding to his responsibility by double-clicking the appropriate item.

The instructors of the course control the flow of time. In order to inform the participants about the current state of ViFE, they have to show interim results of the course, e. g. using a spreadsheet that holds data like turnover, profit, expenses, outstanding debts, and number of sold products. When an order is placed by the client, it is processed in the virtual factory according to the action chain model (see [Toppinen, 1997]), requiring tasks of the participants according to the following departments [Pulkrab, 1997]:

- **Sales department**: This department is responsible for accepting orders or inquiries by the customer and sending order confirmations or offers to him. Furthermore, the staff is negotiating on prices and terms of payment and delivery with the client.
- **Production planning department**: Incoming orders are scheduled by the staff of this department.
- **Purchasing department**: The participants have to watch over the stock, so neither overflow nor underflow of semi-finished parts or products may occur. If necessary, they have to order new semi-finished parts at suppliers and negotiate about their price. When semi-finished products arrive, they have to put them onto stock.
- **Delivery & invoice department**: The staff of this department is responsible for delivery of the finished products, when they leave the manufacturing department. Further they have to invoice the delivery at the customer and to take care that the clients pay in time.
- **Production department**: In this department the orders are transferred from production planning to manufacturing.
- **Manufacturing department**: The participants have to produce the right products in the right quality and amount. For the original installation of VIFTOO in Varkaus, Finland, a small manufacturing line was available, consisting of a one-hand and two-platform robot, one CNC machine, and a large wall unit. This is the part that is substituted by the result of the synthesis of VIFTOO and the SCARA Modeling and Simulation System (see Ch. 13.6.4).

The participants have to take care the products are delivered correctly and in time, and they have to guarantee the financial well-being of the virtual factory. When simulation time is speeded up 20 times to normal time, there can be simulated one working week, i. e. 40 working hours, within a course taking two hours. At the end of the course the quality of the teamwork can be measured by the amount of money the participants have earned for the virtual factory.

The SCARA System for Production Programming and Simulation

SCARA, a system for modeling, programming, and simulation of the SCARA (Selective Compliance Assembly Robot Arm) robot, has been developed at J. Kepler University [Dämon, 1995] and consists of four main modules:

1. The primitive modeling module provides the efficient creation of basic solid components, like boxes, cylinders, polyhedra etc.
2. The object modeling module allows the combination of such primitives into complex composed objects in order to model stationary objects of the real world, like basements, pillars, housings of machines etc.

3. The cell modeling module is used for designing the layout of a cell using the predefined composed objects. With this module the stationary characteristics of a work cell can be completely specified.

4. With the programming module the objects comprising a kinematic behavior (robots, machines, conveyors, etc.) can be programmed. Additionally, simulation runs can be performed using this module.

An example for a simple manufacturing cell that has been modeled using SCARA is shown in Fig. 13.8.

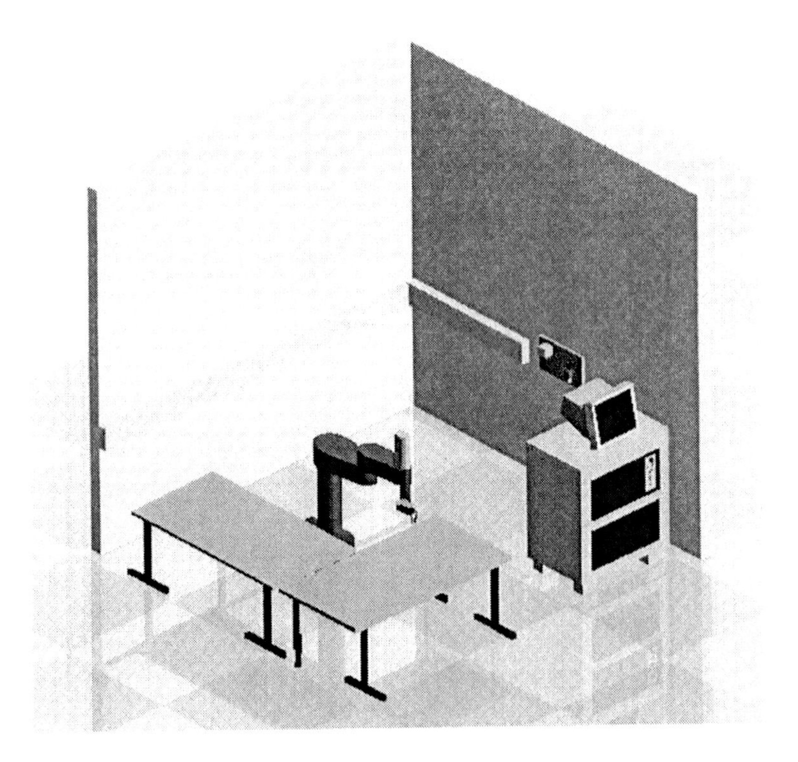

Figure 13.8: Sample cell modeled with SCARA

13.6.3 Migrating from the Existing Components to ViFE

The VIFTOO prototype has been developed using the Microsoft Visual BASIC 3.0 programming environment. The program has been built as simple as possible in order that its use can be adapted quickly. A task of the participants of the courses that were held in Finland during 1996 and early 1997 was also to test the prototype and to give comments and development hints. In this sense the developers of the existing program were not only the programmers, but also the users. Therefore the focus was not put on the quality of the prototype, but on functionality, adaptability, and a short development time.

In order to get a first prototype for an integrated virtual factory system, the VIFTOO production planning system and the SCARA production simulation system have been coupled in the frame of diploma projects at Upper Austrian Polytechnic University, see [Kern, 1997] and [Pulkrab, 1997].

Since the implementation environment between the two systems is very different (Visual BASIC versus C++), a prototypic interface between the two systems has been implemented using files. The main file of the interface controls the scheduling of the production information. Consider the scenario that the planning server and the simulation server are accessed simultaneously by several students performing their corresponding steps of the virtual factory process. The main file is used for storing a reference to a specific file for each task. This file contains information on product identification, lot size, and the priority of the task. By reading and writing appropriate information strings onto the main file, a suitable handshake has been realized for correctly servicing and synchronizing the clients.

13.6.4 Coupling Process Planning and Production Simulation

In order to use SCARA in the frame of ViFE it had to be extended by a communication module in order to be used as a server for production simulation. Additionally, the functionality was extended in order to visualize a simulation not only on a computer screen but also generate a movie file for the simulation that can be transferred to the client stations. There this movie file can be used for visualizing a production. Every time a production parameter or a program is changed, the movie file is automatically generated again and an update is sent to the clients.

The currently modeled manufacturing cell represents the scenario available in hardware at Pohjois-Savo Polytechnic. It consists of a conveyor, a CNC milling machine, a robot, and a wall unit. Programs had to be written for three different products that are currently used as sample products in the hardware plant, too. Additionally, programs had to be written to control the conveyor, to animate the CNC machine, and to organize the general control flow.

Both the communication between the modules and the organization of the database have been redesigned. Since other modules of ViFE, e. g. the market simulator or the quality management module, also hold data in the database (or at least have to access it), there was focused the use of a more powerful and distributed database in order to manage a greater amount of data suitably. Since the ViFE system is also capable of communication via the Internet, a TCP/IP based communication is used for data exchange between the server and its clients.

A sample manufacturing cell is illustrated in Fig. 13.9. For details on the developed ViFE environment, we refer to [Kern, 1997].

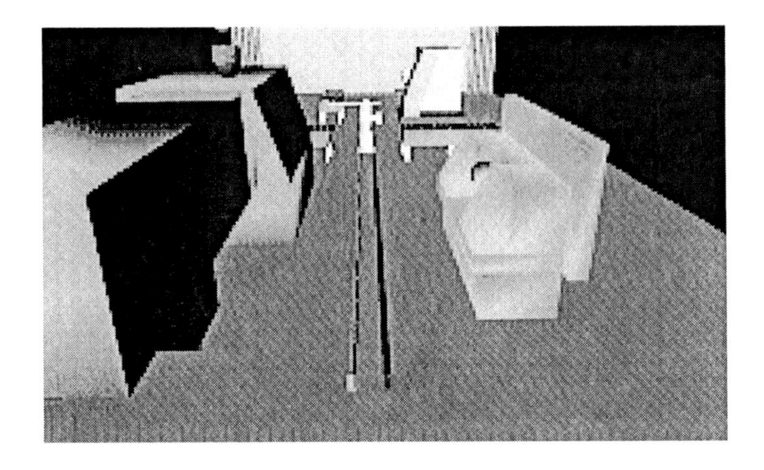

Figure 13.9: Sample ViFE scenario

Literature

[Abeln, 1989] O. Abeln. Referenzmodell für CAD-Systeme (Classification Scheme for CAD Systems; in German). *Informatik Spektrum*, 12:43–46, 1989.

[Abrash, 1992] Michael Abrash. Fast Antialiasing. *Dr. Dobb's Journal*, 17(6):139–146, 1992.

[Agha, 1986] Gul Agha. An Overview of Actor Languages. *SIGPLAN Notices*, 21(10):58 – 67, 1986.

[Ahuja *et al.*, 1980] N. Ahuja, R.T. Chien, R. Yen, and N. Bridwell. Interference Detection and Collision Avoidance Among Three Dimensional Objects. In *Proc. 1st Intl. Conference on Artificial Intelligence*, pages 44–48, Stanford, USA, 1980.

[Anderson, 1978] R.O. Anderson. Detecting and Eliminating Collisions in NC Machining. *Computer Aided Design*, 10(4):231–237, 1978.

[Artmann, 1996] Karl Artmann. *Object-Oriented Engineering Frameworks*. Diploma thesis, RISC-Linz, J. Kepler University, Linz, Austria, Europe, 1996.

[Aukstakalnis and Blatner, 1992] Steve Aukstakalnis and David Blatner. *Silicon Mirage: The Art and Science of Virtual Reality*. Peach Pit Press, 1992.

[Balbach and Sollmann, 1982] J. Balbach and A.A. Sollmann. Automatische Kollisionsüberwachung und Werkzeugeinstellmaß-Ermittlung für die Drehbearbeitung (Automatic Collision Check and Tool Position Generation for Turning; in German). *Industrie Anzeiger*, 104(36):50–52, 1982.

[Baldur and Dube, 1987] R. Baldur and M. Dube. Interference Between Generalized Solid Links. Technical report, Ecole Polytechnique, Montreal, Canada, 1987.

[Barr, 1981] A.H. Barr. Superquadrics and Angle-Preserving Transformations. *IEEE Computer Graphics & Applications*, 1(1):11–23, 1981.

[Barsky and Beatty, 1983] B.A. Barsky and J.C. Beatty. Local Control of Bias and Tension in beta-Splines. *ACM Transactions in Graphics*, 8(2):109–134, April 1983.

[Baumgart, 1974] B.G. Baumgart. *Geometric Modeling for Computer Vision*. PhD thesis, STAN-CS-74-463, Artificial Intelligence Lab, Stanford University, Stanford, USA, 1974.

[Bernatchez, 1995] Marc Bernatchez. Virtual Reality - Introduction to Infographic Technologies. Technical report, Laval University, Quebec City, Canada, 1995.

[Bézier, 1968] P. Bézier. Procédé de définition des courbes et surfaces non mathématiques (Numerical Definition of Non-mathematical Curves and Surfaces; in French). *Automatisme*, 13, 1968.

[Bézier, 1972] P. Bézier. *Numerical Control – Mathematics and Applications*. Wiley-Interscience, New York, USA, 1972. (Translated by A.R. Forrest).

[Bhat and Aziz, 1988] S.P. Bhat and N.M. Aziz. The Implementation of a Simplified CSG Model. *Adv. Eng. Software*, 10(1):26–31, 1988.

[Boyse and Gilchrist, 1981] J.W. Boyse and J.E. Gilchrist. GMSOLID – Interactive Modeling for Design and Analysis of Solids. Technical report GMR-3882, General Motors Research Labs, Warren, MI 48090, USA, November 1981.

[Boyse, 1979] J.W. Boyse. Interference Detection Among Solids and Surfaces. *Communications of the ACM*, 22(1):3–9, 1979.

[Braid *et al.*, 1978] I.C. Braid *et al.* Stepwise Construction of Polyhedra in Geometric Modeling. Technical report 100, CAD Group, Cambridge University, Cambridge, GB, 1978.

[Braid, 1973] I.C. Braid. *Designing with Volumes*. PhD thesis, CAD Group, Cambridge University, Cambridge, GB, 1973.

[Brooks, 1982] R.A. Brooks. Solving the Find–Path Problem by Representing Free Space as Generalized Cones. Technical report A.I. Memo No. 674, Massachusetts Institute of Technology, Artificial Intelligence Laboratory, Cambridge, MA 02139, 1982.

[Brunet and Navazo, 1985] P. Brunet and I. Navazo. Geometric Modeling Using Exact Octree Representation of Polyhedral Objects. In C.E. Vandoni, editor, *Eurographics '85*, pages 159–169. North Holland, 1985.

[Brunner, 1996] S. Brunner. Analyzing Turtle Neck Mobility Using the KISS_ME System. Technical report, RISC-Linz, J. Kepler University Linz, Austria, Europe, 1996.

[Byrne *et al.*, 1993] J. A. Byrne, R. Brandt, and O. Port. The Virtual Corporation. *Business Week*, Feb. 8th:10–13, 1993.

[Cameron, 1984] S. Cameron. *Modelling Solids in Motion.* PhD thesis, Dept. of Artificial Intelligence, University of Edinburgh, UK, 1984.

[Cameron, 1985] S. Cameron. An Implementation of Clash Detection by Four–Dimensional Intersection Tests. Technical report, University of Missouri-Rolla, Graduate Engineering Center, 8001 Natural Bridge Road, St. Louis, MO 63121, USA, 1985.

[Canny, 1984] J. Canny. Collision Detection for Moving Polyhedra. Technical report A.I. Memo No. 806, Massachusetts Institute of Technology, Artificial Intelligence Laboratory, Cambridge, MA 02139, 1984.

[Carlbom *et al.*, 1985] I. Carlbom *et al.* A Hierarchical Data Structure for Representing the Spatial Decomposition of $3D$ Objects. *IEEE Computer Graphics & Applications*, 5(4):24–31, 1985.

[Chang *et al.*, 1997] Tien-Chien Chang, Richard A. Wysk, and Hsu-Pin Wang. *Computer-Aided Manufacturing.* Prentice Hall International Series in Industrial and Systems Engineering, 2nd edition, 1997.

[Chen *et al.*, 1984] Homer H. Chen, Narendra Ahuja, and Thomas S. Huang. Septree Representations of Moving Objects Using Hexagonal Cylindrical Decomposition. *Optical Engineering*, 23(5):531–535, 1984.

[Chin and Feiner, 1989] Norman Chin and Steven Feiner. Near Real-Time Shadow Generation Using BSP Trees. In Richard J. Beach, editor, *Computer Graphics*, volume 23/3, pages 99–106. ACM SIGGRAPH, 1989.

[Cohen, 1990] Jacques Cohen. Constraint Logic Programming Languages. *Communications of the ACM*, 33(7):52–68, 1990.

[Cole and Yap, 1985] Richard Cole and Chee K. Yap. Shape from Probing. Technical report, Robotics Lab, Courant Institute of Mathematical Sciences, New York University, N.Y. 10012, USA, October 1985.

[Collins, 1975] George E. Collins. Quantifier Elimination for Real Closed Fields by Cylindrical Algebraic Decomposition. In *Proc. 2nd GI Conference on Automata Theory and Formal Languages*, pages 134–183, Berlin, 1975. Lecture Notes in Computer Science, Springer Verlag.

[Comba, 1968] P.G. Comba. A Procedure for Detecting Intersections of Three-Dimensional Objects. *Journal of the ACM*, 15(3):354–366, 1968.

[Coons, 1967] S.A. Coons. Surfaces for Computer-Aided Design of Space Forms. Technical report MAC-TR-41, Massachussetts Institute of Technology, 1967.

[Cordonnier *et al.*, 1985] E. Cordonnier *et al.* Creating CSG Modeled Pictures for Ray-Casting Display. In C.E. Vandoni, editor, *Eurographics '85*, pages 171–182. North Holland, 1985.

[Craig, 1986] J.J. Craig. *Introduction to Robotics Mechanics & Control.* Addison Wesley, 1986.

[Craiger, 1999] Philip Craiger. Human Computer Interaction. Technical report, Departments of Computer Science & Informaton Systems Engineering, University of Nebraska at Omaha, Omaha, NE 68182-0459, USA, 1999.

[Cugini, 1988] U. Cugini. The Role of Different Levels of Modeling in CAD Systems. In R.A. Earnshaw, editor, *Theoretical Foundations of Computer Graphics and CAD*, NATO ASI Series, pages 881–898. Springer Verlag, 1988.

[Dämon, 1995] J. Dämon. *Ein System zur Offline-Programmierung des SCARA-Roboters (A System for Offline Programming the SCARA Robot; in German).* Diploma thesis, Department of Systems Engineering and Automation, J. Kepler University, Linz, Austria, Europe, 1995.

[Davidow and Malone, 1992] W. H. Davidow and M. S. Malone. *The Virtual Corporation. Structuring and Revitalizing the Corporation for the 21st Century.* New York, 1992.

[Denavit and Hartenberg, 1954] J. Denavit and R.S. Hartenberg. A Kinematic Notion for Lower-pair Mechanisms Based on Matrices. *Trans. of the ASME, J. of Applied Mechanics*, 22:215–221, 1954.

[Desoyer *et al.*, 1985] Kurt Desoyer, Peter Kopacek, and Inge Troch. *Industrieroboter und Handhabungsgeräte (Industrial Robots and Handling Devices; in German).* Oldenbourg Verlag, München, Wien, 1985.

[Diedenhoven, 1985] H. Diedenhoven. Usable Information Content from CAD Data for NC-Production. *CAE Journal*, 5:58–62,64–65, 1985.

[Dombre *et al.*, 1985] E. Dombre *et al.* Design of a CAD/CAM System for Robotics on a Microcomputer. In O.D. Faugeras and G. Giralt, editors, *Proc. of the Third International Symposium on Robotics Research*, pages 175–184, Cambridge, Massachusetts, 1985. Gonvieux, France, The MIT Press.

[Droy, 1981] J. Droy. Machining Centers Mean Production Profits. *Production Engineering*, June:50–54, 1981.

[Drysdale *et al.*, 1989] Robert L. Drysdale, Robert B. Jerard, Barry Schaudt, and Ken Hauck. Discrete Simulation of NC Machining. *Algorithmica*, 4:33–60, 1989.

[Duelen *et al.*, 1990] G. Duelen, H.-D. Stahlmann, M. Imam, O. Reimann, and A. Strippel. Steuerungserweiterung zur Online-Kollisionsüberwachung (Control Extension for Online Collision Monitoring; in German). *ZwF CIM – Zeitschrift für wirtschaftliche Fertigung und Automatisierung*, 85(8):427–430, 1990.

[Eastman and Weiler, 1979] C.M. Eastman and K. Weiler. Geometric Modeling Using the Euler Operators. Technical report no. 78, Inst. Physical Planning, Carnegie-Mellon University, Pittsburgh, PA, USA, February 1979.

[Eigner *et al.*, 1991] Martina Eigner, Christine Hiller, Stephan Schindewolf, and Matthias Schmich. *Engineering Database: Strategische Komponente in CIM-Konzepten (Engineering database: Strategic Component in CIM Concepts; in German).* Carl Hanser Verlag, München Wien, 1991.

[Ejiri *et al.*, 1972] M. Ejiri *et al.* A Prototype Intelligent Robot that Assembles Objects from Plane Drawings. *IEEE Transactions on Computers*, C-21(2):199–207, 1972.

[Elgabry, 1984] A.K. Elgabry. Integral Link Between Geometric Modeling and CAM Applications. In *Proc. CASA/SME AUTOFACT 6 Conf.*, pages 6.1–6.12, Anaheim, California, USA, 1984.

[Encarnaçao and Schlechtendahl, 1983] J.L. Encarnaç
ao and E.G. Schlechtendahl. *Computer Aided Design. Fundamentals and System Architectures.* Symbolic Computation – Computer Graphics. Springer Verlag, 1983.

[Engelberger, 1980] J.F. Engelberger. *Robotics in Practice.* IFS Publications Ltd., Kempston, England, 1980.

[Eversheim and Holz, 1982] W. Eversheim and B. Holz. Computer Aided Programming of NC-Machine Tools Using the System AUTAP-NC. *Annals of the CIRP*, 31(1):323–327, 1982.

[Eversheim *et al.*, 1988] W. Eversheim *et al.* Kollisionskontrolle für offline erstellte Industrieroboter–Programme (Collision Checks for Offline Industrial Robot Programming; in German). *VDI-Z – Entwicklung, Konstruktion, Produktion*, 130(1):63–68, 1988.

[Fang *et al.*, 1991] Sheng Fang, Fan Xien, and Sun Zengqi. A Simulation System for Dynamic Analysis and Control Research of Robotic Manipulators. In *Proc. IFAC Symposium on Robot Control (SYROCO'91)*, pages 441–446, Vienna, Austria, Sep. 16-18, 1991.

[Flanagan, 1998] David Flanagan. *Java in a Nutshell.* O'Reilly, 2nd edition, 1998.

[Foley *et al.*, 1996] James D. Foley, Andries van Dam, Steven K. Feiner, and John F. Hughes. *Computer Graphics : Principles and Practice, Second Edition in C.* Addison-Wesley, 1996.

[Fuh *et al.*, 1985] K.-H. Fuh *et al.* Computer Aided Design for Different Drill Geometries by the Quadratic Surface Model. In H.-J. Bullinger and H.-J. Warnecke, editors, *Toward the Factory of the Future*, pages 425–430, Germany, 1985. Fraunhofer Institut for Industrial Engineering, Springer Verlag.

[Gardan, 1983] Y. Gardan. A System for the Interactive Description of Parameterized Elements. In T.M.R. Ellis and O.I. Semenkov, editors, *Advances in CAD/CAM*, pages 159–167. IFIP/IFAC Proceedings PROLAMAT '82, North Holland, 1983.

[Geiger, 1978] M. Geiger. Blech als Konstruktionswerkstoff im Werkzeugmaschinenbau (Using Sheet Metal for NC Machine Construction; in German). *Werkstattstechnik – Zeitschrift für industrielle Fertigung*, 68:193–201, 1978.

[Gerke, 1985] W. Gerke. Die dynamische Programmierung zur Planung kürzester kollisionsfreier Bahnen für Industrieroboter (The Dynamic Programming for the Planning of Shortest Collision Free Paths for Industrial Robots; in German). *Robotersysteme*, 1(1):43–52, 1985.

[Ghee, 1995] Steve Ghee. dVS - a Distributed VR System Infrastructure. In *Proc. SIGGRAPH '95.* Los Angeles, California, USA, ACM Press, 1995.

[Gödde, 1990] B. Gödde. Die WOP-Philosophie am Beispiel Fräsen (The WOP Philosophy Taking Milling as an Example; in German). Technical report 90-2335, Robert Bosch GmbH, Stuttgart, Germany, 1990.

[Goertz, 1963] R.C. Goertz. *Manipulators Used for Handling Radioactive Materials.* Human Factors in Technology, chapter 27. McGraw-Hill, 1963.

[Goldman, 1987] Ronald N. Goldman. The Role of Surfaces in Solid Modeling. In G.E. Farin, editor, *Geometric Modeling: Algorithms and New Trends*, pages 69–90. SIAM, 1987.

[Gordon, 1969] W.J. Gordon. Spline-Blended Surface Interpolation Through Curve Networks. *J. Math. Mech.*, 18:931–952, 1969.

[Graiser, 1983] H.N. Graiser. CAD/Robotics Capability Helps Realize CIM. *IF*, pages 26–30, August 1983.

[Greenhalgh, 1994] Chris Greenhalgh. An Experimental Implementation of the Spatial Model. In *Proc. 6th ERCIM Workshop.* Swedish Institute of Computer Science (SICS), Stockholm-Kista, June 1-3, 1994.

[Groover and Zimmers, 1984] M.P. Groover and E.W. Zimmers. *CAD/CAM: Computer-Aided Design and Manufacturing.* Prentice-Hall, 1984.

[Hajek and Harmanec, 1992] Petr Hajek and David Harmanec. On Belief Functions. Technical report, Institute of Computer and Information Science, Czechoslovak Academy of Sciences, 182 07 Prague, Czechoslovakia, 1992.

[Hametner, 1996] Andreas Hametner. *A Programming Environment for an Object-Oriented Engineering Framework.* Diploma thesis, RISC-Linz, J. Kepler University, Linz, Austria, Europe, 1996.

[Hanrahan, 1982] P. Hanrahan. Creating Volume Models from Edge-Vertex Graphs. *ACM SIGGRAPH Computer Graphics*, 16(3):77–84, July 1982.

[Hase, 1997] Hans-Lothar Hase. *Dynamische Virtuelle Welten mit VRML 2.0.* dpunkt, 1st edition, 1997.

[Heintze *et al.*, 1988] Nevin Heintze, Spiro Michaylov, and Peter Stuckey. Constraint Logic Programming and Some Electrical Engineering Problems. Technical report, Department of Computer Science, Monash University, Clayton 3168, Victoria, Australia, 1988.

[Heinzelreiter and Mayr, 1990] Johann Heinzelreiter and Herwig Mayr. Solid Modeling of Dynamic Objects Using Spatial Enumeration Techniques - A Survey. Lecture Notes no. 90-23.0 (RISC-Linz Series), RISC-Linz, J. Kepler University, Linz, Austria, Europe, 1990.

[Heinzelreiter, 1990] Johann Heinzelreiter. Documentation of the SAVE 3D Dynamic Modeling Scheme. Technical report, RISC-Linz, J. Kepler University, Linz, Austria, Europe, 1990.

[Hellwagner, 1987] Hermann Hellwagner. Computer-Graphik (Computer Graphics; in German). Lecture notes, Systems Sciences Dept., Univ. Linz, Austria, 1987.

[Hersh, 1988] Jay S. Hersh. Tools for Particle Based Geometric Modeling. Technical report, Rensselaer Polytechnic Institute, Troy, New York, 1988.

[Hintenaus, 1987] Peter Hintenaus. The Inverse Kinematics System (Installation Guide, User's Manual, Program Documentation). Technical report no. 87-18.0 (RISC-Linz Series), RISC-Linz, J. Kepler University, Linz, Austria, Europe, 1987.

[Hoffman and Hebert, 1986] R.M. Hoffman and M. Hebert. Applications of High-performance Graphics Workstations in Robotic Simulation. In H. Van Brussel, editor, *Proc. 16th International Symposium on Industrial Robots, 8th International Conference on Industrial Robot Technology*, pages 775–784, Berlin Heidelberg New York Tokyo, 1986. September 30 – October 2, 1986, Brussels, Belgium, IFS (Publications) Ltd, UK, Springer Verlag.

[Hoffmann, 1989] Christoph M. Hoffmann. *Geometric and Solid Modeling.* Morgan Kaufmann Publishers, Inc., San Mateo, California 94403, 1989.

[Hohn, 1976] R.E. Hohn. Application Flexibility of a Computer Controlled Industrial Robot. Technical report MR 76-603, SME Technical Paper, Society of Mechanical Engineers, 1976.

[Holzer and Pirngruber, 1997] Helmut Holzer and Alexander Pirngruber. The CAVE. A Virtual Reality enviroment. *http://www.ce.uni-linz.ac.at/courses/ss97/BASeminar97/holzer.htm*, 1997. [Accessed April 6, 2000].

[Hook and Tiller, 1989] D. Hook and W. Tiller. Boolean Operations on *3D* Objects Defined as Collections of Trimmed Surfaces. In *Proc. Theory and Practice of Geometric Modeling*, October 1988, Blaubeuren, Germany, Springer Verlag, 1989.

[Hopcroft *et al.*, 1983] J. E. Hopcroft, J. T. Schwarz, and M. Sharir. Efficient Detection of Intersections among Spheres. *International Journal of Robotics Research*, 2(4):77–80, 1983.

[Hui and Yi, 1988] Z. Hui and Q. Yi. Solving the Workspace Problem of Robot on the Basis of Graphical and Analytical Methods. In *Proceedings ICEGDG*, pages 331–336. Vienna, Austria, 1988.

[InterSense, 1999] InterSense. The InterSense Homepage. *http://www.isense.com*, 1999. [Accessed April 6, 2000].

[Isdale, 1993] Jerry Isdale. What Is Virtual Reality? A Homebrew Introduction and Information Resource List. Technical report, Isdale Engineering, California, USA, 1993.

[Jaissle, 1977] H.-U. Jaissle. Im Sondermaschinenbau muß standardisiert werden (Standardization is Necessary in the Construction of Special Purpose NC Machines; in German). *Moderne Fertigung*, 11:43–44, 1977.

[Jenewein, 1999] Martin Jenewein. *Modellierung interaktiver VRML-Welten im CAVE (Modeling of Interactive VRML Worlds in the CAVE; in German).* Diploma thesis, Department of Software Engineering, Upper Austrian Polytechnic University, Hagenberg, Austria, Europe, 1999.

[Jerard *et al.*, 1989] Robert B. Jerard, Robert L. Drysdale, Kenneth Hauck, Barry Schaudt, John Magewick, and Ken Hauck. Methods for Dedecting Errors in Numerically Controlled Machining of Sculptured Surfaces. *IEEE Computer Graphics & Applications*, 9(1):26–39, 1989.

[Karppanen, 1996] Erkki Karppanen. Virtual Factory for Education. Proposal for EU Research Project (2nd Year). Technical report, W. Ahlström Institute of Technology, Varkaus, Finland, 1996.

[Kern, 1997] Gerald Kern. *Migration eines Modellier- und Simulationssystems für Produktionsabläufe zu einem Client/Server-System für Virtual-Factory-Anwendungen (Migration of a Modeling and Simulation System for Production Process to a Client/Server System for Virtual Factories; in German).* Diploma thesis, Department of Software Engineering, Upper Austrian Polytechnic University, Hagenberg, Austria, Europe, 1997.

[Khatib and LeMaitre, 1978] O. Khatib and J.-F. LeMaitre. Dynamic Control of Manipulators Operating in a Complex Environment. In *Proc.* 3rd *CISM-IFTOMM Symp. on the Theory and Practice of Robots and Manipulators*, pages 267–282, Udine, Italy, 1978.

[Kochan, 1986] D. Kochan, editor. *CAM. Developments in Computer Integrated Manufacturing.* IFIP State of the Art report. Springer Verlag, Berlin-Heidelberg, 1986.

[Kondo and Kimura, 1988] K. Kondo and F. Kimura. Collision Avoidance Using a Free Space Enumeration Method Based on Grid Expansion. *Advanced Robotics*, 3, 1988.

[Kowal, 1992] J.A. Kowal. *Behavior Models: Specifying User's Expectations.* Prentice Hall, Englewood Cliffs, New Jersey, USA, 1992.

[Krautter and Steinert, 1989] Wolfgang Krautter and Michael Steinert. A Knowledge Representation for Model-Based Reasoning Using Prolog III. Technical report P1219 (1106), Robert Bosch GmbH, Karlsruhe, Germany, 1989.

[Kruger, 1983] Myron Kruger. *Artificial Reality.* Addison-Wesley, 1983.

[Latombe, 1991] Jean-Claude Latombe. Motion Planning: Theory and Applications. In G. Doumeingts, J. Browne, and M. Tomljanovich, editors, *Proc. Computer Applications in Production and Engineering CAPE'91*, pages 23–32. International Federation for Information Processing (IFIP), Elsevier Science Publishers B.V., 1991.

[Laura *et al.*, 1996] Lemay. Laura, Kelly Murdock, and Justin Couch. *3D Graphics and VRML 2.* Sams.net Publishing, 1996.

[Law and Kelton, 1982] A.M. Law and W.D. Kelton. *Simulation Modeling and Analysis.* McGraw-Hill, New York, 1982.

[Lehmann, 1989] Clemens M. Lehmann. *Wissensbasierte Unterstützung von Konstruktionsprozessen (Knowledge Based Support of Design Processes; in German).* Carl Hanser Verlag, München Wien, 1989.

[Liegeois *et al.*, 1984] A. Liegeois *et al.* Programming, Simulating and Evaluating Robot Actions. In H. Hanafusa and H. Inoue, editors, *Proc. of the Second International Symposium on Robotics Research*, pages 411–415, Cambridge, Massachusetts, 1984. Kyoto, Japan, The MIT Press.

[Loeffler and Anderson, 1994] C. E. Loeffler and T. Anderson. *The Virtual Reality Casebook.* International Thomson Publishing, New York, USA, 1994.

[Lozano-Perez and Wesley, 1979] T. Lozano-Perez and M.A. Wesley. An Algorithm for Planning Collision-Free Paths Among Polyhedral Obstacles. *Communications of the ACM*, 22(10):560–570, 1979.

[Lozano-Perez, 1977] T. Lozano-Perez. LAMA: A Language for Automatic Mechanical Assembly. In *Proc. 5th Intl. Joint Conference on Artificial Intelligence*. Massachusetts Institute of Technology, 1977.

[Lozano-Perez, 1980a] T. Lozano-Perez. Automatic Planning of Manipulator Transfer Movements. Technical report A.I. Memo No. 606, Massachusetts Institute of Technology, Artificial Intelligence Laboratory, Cambridge, MA 02139, 1980.

[Lozano-Perez, 1980b] T. Lozano-Perez. Spatial Planning: A Configuration Space Approach. Technical report A.I. Memo No. 605, Massachusetts Institute of Technology, Artificial Intelligence Laboratory, Cambridge, MA 02139, 1980.

[Lumelsky, 1986] V.J. Lumelsky. Continuous Motion Planning in Unknown Environment for a 3D Cartesian Robot Arm. In *Proc. IEEE Int. Conf. on Robotics and Automation*, volume 2, 1986.

[Lutz, 1988] M. Lutz. Methode zur Sicherung der geometrischen Kondition bei der Punktlage–Bestimmung (Method for Saving the Geometric Condition during Point Location; in German). *ZwF CIM – Zeitschrift für wirtschaftliche Fertigung und Automatisierung*, 83(5):247–252, 1988.

[Macedonia *et al.*, 1995] Michael R. Macedonia, Michael J. Zyda, David R. Pratt, Donald P. Brutzman, and Paul T. Barham. Exploiting Reality with Multicast Groups: A Network Architecture for Large-scale Virtual Environments. *IEEE Computer Graphics and Applications*, 95(9):38 – 45, 1995.

[Mäntylä and Sulonen, 1982] M. Mäntylä and R. Sulonen. GWB: A Solid Modeler with Euler Operators. *IEEE Computer Graphics & Applications*, 2(7):17–31, September 1982.

[Marko, 1996] Monika Marko. A Hierarchical Solid Modeler for Simulation in Virtual Worlds Using Object-Oriented Techniques. Diploma thesis, RISC-Linz, Dept. of Mathematics, Johannes Kepler University, Linz, Austria, 1996.

[Markowsky and Wesley, 1980] G. Markowsky and M.A. Wesley. Fleshing Out Wire-Frames. *IBM*, 24(5):582–597, September 1980.

[Mayr and Heinzelreiter, 1991a] Herwig Mayr and Johann Heinzelreiter. A Solid Modeler for Dynamic Objects Using the Dexel Representation. In V. Hubka, editor, *Proc. International Conference on Engineering Design*, pages 1082–1085. Zürich, Switzerland, August 27–29, 1991, Edition HEURISTA, 1991.

[Mayr and Heinzelreiter, 1991b] Herwig Mayr and Johann Heinzelreiter. Modeling and Simulation of the Robotics/NC Machining Process Using a Spatial Enumeration Representation. In Paolo Dario, editor, *Proc. 5th Intl. Conference on Advanced Robotics, ICAR91*, pages 1594–1597, Piscataway, NJ, USA, 1991. Pisa, Italy, June 20-22, IEEE.

[Mayr and Oberreiter, 1986] Gerhard Mayr and Walter Oberreiter. RobLan–Sprachbeschreibung (RobLan Language Description; in German). Technical report, VOEST ALPINE AG, Linz, Austria, 1986.

[Mayr and Öllinger, 1991] Herwig Mayr and Hermann Öllinger. *S M A R T – Simulation of Manufacturing and Robot Tasks*. In Dieter W. Wloka, ed., Robotersimulation, pages 153–186. Springer Verlag, Berlin Heidelberg New York, 1991.

[Mayr and Stifter, 1989] Herwig Mayr and Sabine Stifter. Offline Generation of Error-free Robot/NC Code Using Simulation and Automatic Programming Techniques. In G. Halevi, editor, *Proc. International Conference on CAD/CAM and AMT*, Binyanei Ha'ooma, Jerusalem, Israel, Dec. 11 – 14, 1989. North Holland.

[Mayr *et al.*, 1989] Herwig Mayr, Martin Held, and Hermann Öllinger. SMART. A Universal System for the Simulation of Machining and Robot Tasks. In *Proc. Computer Applications in Production and Engineering (CAPE'89)*, pages 809–816, Tokyo, Japan, 1989. North Holland, 1989.

[Mayr, 1987] Herwig Mayr. *F A A S T – Fully Automatic Area Stamping*. Diploma thesis, Technical report no. 87-43.0 (RISC-Linz Series), RISC-Linz, J. Kepler University, Linz, Austria, Europe, 1987.

[Mayr, 1990] Herwig Mayr. Highly-Efficient Collision Checking for Robotics/NC Using Linear Programming Techniques. In G. Feichtinger, editor, *Proc. Intl. Conference on Operations Research*. Vienna, Austria, August 28–31, 1990.

[Mayr, 1991a] Herwig Mayr. NC Machines Meet Robots – Towards a Common Robotics/NC Standard. In G. Doumeingts, J. Browne, and M. Tomljanovich, editors, *Proc. 4th CAPE – Computer Applications in Production and Engineering*, pages 555–562. Bordeaux, France, Sept. 10–12, 1991, North Holland, Amsterdam, 1991.

[Mayr, 1991b] Herwig Mayr. *Real-Time Dynamic Collision Checking Integrated into Graphic Manufacturing Simulation*. PhD thesis, Technical report no. 91-61.0 (RISC-Linz Series), RISC-Linz, J. Kepler University, Linz, Austria, Europe, 1991.

[Mayr, 1991c] Herwig Mayr. The Concept of NC Machine Geometries Within SAVE. Technical report no. 91-42.0 (RISC-Linz Series), RISC-Linz, J. Kepler University, Linz, Austria, Europe, 1991.

[Mayr, 1997a] Herwig Mayr. GEM: A Generic Engineering Framework for Mechanical Engineering Based upon Meta Models. In R. Moreno-Diaz and F. Pichler, editors, *Proc. EUROCAST'97.* Februar 23 - 28, Las Palmas de Gran Canaria, Spain, 1997.

[Mayr, 1997b] Herwig Mayr. Using Software Sensors for Migrating From Classical Simulation Systems Towards Virtual Worlds. In *Proc. Engineering of Computer Based Systems.* March 24-28, Monterey, CA, USA, IEEE, 1997.

[Meagher, 1980] D.J. Meagher. Octree Encoding: A New Technique for the Representation, Manipulation, and Display of Arbitrary Three-Dimensional Objects by Computer. Technical report IPL-TR-80-111, Image Processing Lab, Rensselaer Polytechnic Institute, Troy, NY 12181, USA, October 1980.

[Miller, 1985] Richard K. Miller. Manufacturing Simulation. New Tool for Robotics, FMS, and Industrial Process Design. Technical report, SEAI Technical Publications, Madison, GA 30650, USA, 1985.

[Mills, 1985] R.B. Mills. Robots Move into Manufacturing Cells. *CAE – Computer Aided Engineering,* 10:40–48, October 1985.

[Moore, 1980] R.E. Moore. Interval Methods for Nonlinear Systems. *Computing,* Suppl. 2:113–120, 1980.

[Mortenson, 1985] M.E. Mortenson. *Geometric Modeling.* John Wiley and Sons, 1985.

[Myers and Agin, 1983] J.K. Myers and G.J. Agin. A Supervisory Collision Avoidance System for Robot Controllers. *Robotics World,* 1(1):225–232, 1983.

[Myers, 1981] J.K. Myers. A Supervisory Collision Avoidance System for Robot Controllers. Master's thesis, Electrical Engineering Department, Carnegie Mellon University, 1981.

[Myers, 1985] J.K. Myers. A Robotic Simulator with Collision Detection: RCODE. In *Proc.* 1st *Annual Workshop on Robotics and Expert Systems,* pages 205–213. ISA, June 1985.

[Nahrstedt and Steinmetz, 1995] Klara Nahrstedt and Ralf Steinmetz. Resource Management in Networked Multimedia Systems). *IEEE Computer,* 28(5):52 - 63, 1995.

[Navazo *et al.*, 1986] I. Navazo *et al.* A Geometric Modeler Based on the Exact Octree Representation of Polyhedra. *Computer Graphics Forum,* 5(2):89–104, June 1986.

[Neider *et al.*, 1993] J. Neider, T. Davis, and M. Woo. *OpenGL Programming Guide.* Silicon Graphics, 1993.

[Nutbourne and Martin, 1988] A.W. Nutbourne and R.R. Martin, editors. *Differential Geometry Applied to Curve and Surface Design.* Vol. 1: Foundations. ASIN: 013211822X, 1988.

[Ocken *et al.*, 1987] S. Ocken *et al.* Precise Implementation of CAD Primitives Using Rational Parametrizations of Standard Surfaces. In J.T. Schwartz *et al.*, editor, *Planning, Geometry and Complexity of Robot Motion,* Ablex Series in Artificial Intelligence, pages 245–266. Ablex Publishers, Norwood, New Jersey, 1987.

[Ozaki *et al.*, 1984] H. Ozaki *et al. On the Collision Free Movement of a Manipulator,* pages 189–200. Advanced Software in Robotics. A. Danthine and M. Geradin (eds.), Elsevier Science Publishers B.V. (North–Holland), 1984.

[Paul, 1981] R.P. Paul. *Robot Manipulators: Mathematics, Programming, and Control.* MIT Press Series in Artificial Intelligence. Massachusetts Institute of Technology, 1981.

[Peneder, 1996] Leopold Peneder. KISS_ME - Kinematic Interactive Simulation System and Modeling Environment. Diploma thesis, RISC-Linz, Dept. of Mathematics, Johannes Kepler University, Linz, Austria, 1996.

[Peterson, 1984] D. Peterson. Halfspace Representation of Extrusions, Solids of Revolution, and Pyramids. Technical report SAND84-0572, SANDIA National Laboratories, NM, USA, 1984.

[Pichler and Schwärzel, 1992] F. Pichler and H. Schwärzel. *CAST Methods in Modeling.* Springer Verlag, 1992.

[Pieper, 1968] D.L. Pieper. *The Kinematics of Manipulators Under Computer Control.* PhD thesis, Stanford Artificial Intelligence Lab, Stanford University, USA, 1968.

[Pilland, 1988] U. Pilland. Echtzeit-Kollisionsschutzsysteme für NC-Drehmaschinen (Systems for Real-time Collision Avoidance at NC Lathes; in German). *Werkstattstechnik – Zeitschrift für industrielle Fertigung,* 78:509–514, 1988.

[Pinkler and Simon, 1976] G. Pinkler and V. Simon. A General Dialogue System for Interactive Graphic Programming of NC Machines and CAD Systems. In *Proc.* 3rd *IFIP-IFAC Intl. Conference on Programming Languages for Numerically Controlled Machine Tools*, Stirling, 1976.

[Pomberger and Blaschek, 1996] Gustav Pomberger and Günther Blaschek. *Grundlagen des Software Engineering: Prototyping und objektorientierte Software-Entwicklung (Foundations of Software Engineering: Prototyping and Object-Oriented Software Development; in German).* Carl Hanser Verlag, München Wien, second edition, 1996.

[Potthast *et al.*, 1988] A. Potthast, S.H. Kwok, and Y.-S. Lim. Rechnerische Kollisionskontrolle mit einem dynamischen 3D-Simulationssystem (Algorithmic Collision Detection in a Dynamic 3D Simulation System; in German). *ZwF CIM - Zeitschrift für wirtschaftliche Fertigung und Automatisierung*, 83(3):153–157, 1988.

[Pratt, 1987] M.J. Pratt. Surface Modeling. In J. Rooney and P. Steadman, editors, *Principles of Computer-aided Design*, pages 107–116. Pitman Publishing, 1987.

[Preparata and Shamos, 1993] F.P. Preparata and M.I. Shamos. *Computational Geometry - An Introduction.* Texts and Monographs in Computer Science. Springer Verlag, New York, 2nd edition, 1993.

[Pritschow and Kayser, 1987] G. Pritschow and K.-H. Kayser. 3D-Echtzeitkollisionsüberwachung an Fertigungseinrichtungen (3D Real Time Collision Checking for Manufacturing Cells; in German). *Werkstattstechnik - Zeitschrift f. industrielle Fertigung*, 77:201–205, 1987.

[Pritsker and Alan, 1984] A. Pritsker and B. Alan. *Introduction to Simulation and SLAM II.* Halsted Press, New York and Systems Publishing Corporation, West Lafayette, 1984.

[Pulkrab, 1997] Jürgen Pulkrab. *Synthesis of a Process Planning Module based on the Order Processing Module for a Virtual Factory.* Diploma thesis, Department of Software Engineering, Upper Austrian Polytechnic University, Hagenberg, Austria, Europe, 1997.

[Rahmacher and Heßelmann, 1983] K. Rahmacher and J. Heßelmann. Erfahrungen bei der Entwicklung und Anwendung eines graphischen Prozeß-Simulationssystems (Experiences of the Development and Application of a Graphic Process Simulation System; in German). *ZwF CIM - Zeitschrift für wirtschaftliche Fertigung und Automatisierung*, 78(6):276–279, 1983.

[Ránky and Ho, 1985] P.G. Ránky and C.Y. Ho. *Robot Modeling.* Springer Verlag, 1985.

[Red, 1983] W.E. Red. Minimum Distances for Robot Task Simulation. *Robotica*, 1:231 – 238, 1983.

[Rembold and Dillmann, 1986] U. Rembold and R. Dillmann. *Computer-Aided Design and Manufacturing.* Symbolic Computation - Computer Graphics. Springer Verlag, 1986.

[Rennau and Schnitzler, 1984] W. Rennau and M. Schnitzler. Ein graphisches Roboter–Simulationssystem (A Graphical Robot Simulation System; in German). *ZwF CIM - Zeitschrift für wirtschaftliche Fertigung und Automatisierung*, 79(9):409–412, 1984.

[Requicha and Voelcker, 1977] A.A.G. Requicha and H.B. Voelcker. Constructive Solid Geometry. Technical report no. 25, Production Automation Project, University of Rochester, Rochester, NY 14627, USA, 1977.

[Requicha and Voelcker, 1979] A.A.G. Requicha and H.B. Voelcker. Geometric Modeling of Mechanical Parts and Machining Processes. In *COMPCONTROL'79*, Sopron, Hungary, November 1979.

[Requicha, 1977] A.A.G. Requicha. Mathematical Models of Rigid Solid Objects. Technical report no. 28, Production Automation Project, University of Rochester, Rochester, NY 14627, USA, 1977.

[Requicha, 1980] A.A.G. Requicha. Representations for Rigid Solids: Theory, Methods and Systems. *ACM Computing Surveys*, 12(4):437–464, 1980.

[Riesenfeld, 1973a] R. Riesenfeld. *Applications of B-Spline Approximation to Geometric Problems of Computer-Aided Design.* PhD thesis, Syracuse University, Syracuse, NY, USA, 1973.

[Riesenfeld, 1973b] R. Riesenfeld. Applications of B-Spline Approximation to Geometric Problems of Computer-Aided Design. Technical report UTEC-Csc-73-126, Dept. of CS, University of Utah, USA, 1973.

[Roberts, 1965] L.G. Roberts. Homogeneous Matrix Representation and Manipulation of N-Dimensional Constructs. Technical report MS 1045, Lincoln Laboratory, Massachusetts Institute of Technology, USA, 1965.

[Roehl, 1995] Bernie Roehl. Distributed Virtual Reality - An Overview. Technical report, http://sunee.uwaterloo.ca/~broehl/distrib.html, June 1995. [Accessed July 4, 2000].

[Roider and Stifter, 1987] Bernhard Roider and Sabine Stifter. Collision of Convex Objects. In J. Davenport, editor, *Proc. EUROCAL'87.* Leipzig, Germany, June 2-5, 1987, Springer Verlag, 1987.

[Rooney, 1987a] J. Rooney. Geometry in Motion. In J. Rooney and P. Steadman, editors, *Principles of Computer-aided Design*, pages 285–295. Pitman Publishing, 1987.

[Rooney, 1987b] J. Rooney. Representing Objects. In J. Rooney and P. Steadman, editors, *Principles of Computer-aided Design*, pages 13–37. Pitman Publishing, 1987.

[Rosenberg, 1972] J. Rosenberg. A History of Numerical Control 1949–1972: The Technical Development, Transfer to Industry, and Assimilation. Technical report ISI-RR-72-3, U.S.C. Information Sciences Institute, Marina del Rey, California, USA, 1972.

[Rozenblit and Hu, 1992] Jerzy W. Rozenblit and Jhyfang Hu. Integrated Knowledge Representation and Management in Simulation-based Design Generation. *Mathematics and Computers in Simulation*, 34:261–282, 1992.

[Samet and Webber, 1988] H. Samet and R.E. Webber. Hierarchical Data Structures and Algorithms for Computer Graphics. Part I: Fundamentals. *IEEE Computer Graphics & Applications*, 8(3):48–68, 1988.

[Samet, 1984] Hanan Samet. The Quadtree and Related Hierarchical Data Structures. *Computing Surveys*, 16(2):187–260, 1984.

[Sarraga, 1983] R.F. Sarraga. Algebraic Methods for Intersections of Quadric Surfaces in GMSOLID. *CVG*, 22:222–238, 1983.

[Sautter, 1987] R. Sautter. *Numerische Steuerungen für Werkzeugmaschinen (Numerical Control Systems of Machine Tools; in German).* Vogel-Buchverlag, Würzburg, Germany, 1987.

[Schade and Schade, 1990] Bernd Schade and Klaus-Günther Schade. Simulation bei der NC-Programmierung (Simulation for NC Programming; in German) . *ZwF CIM - Zeitschrift für wirtschaftliche Fertigung und Automatisierung*, 85(3):CA 28 - CA 31, 1990.

[Schmiedmayer, 1987] H.B. Schmiedmayer. *Simulation von Roboterbewegungen in graphischer Darstellung unter Berücksichtigung eines Kollisionstests (Graphic Simulation of Robot Motions Under Consideration of Collision Detection; in German).* Diploma thesis, Institut für Wasserkraftmaschinen und Pumpen, Abteilung für Regelungstechnik, Technische Universität Wien, Vienna, Austria, 1987.

[Schöling and Reles, 1983] H. Schöling and T. Reles. Offline–Kollisionskontrolle bei Industrierobotern (Offline Collision Detection for Industrial Robots; in German). *VDI-Z - Entwicklung, Konstruktion, Produktion*, 125(17):647–652, 1983.

[Schräder, 1996] A. Schräder. *Management virtueller Unternehmungen (Management of Virtual Enterprises; in German).* Campus Verlag, Frankfurt New York, 1996.

[Schwartz and Sharir, 1983] J.T. Schwartz and M. Sharir. On the "Piano Movers' " Problem, Parts I, II and III. *Communications on Pure and Applied Mathematics*, 36:345–398, 1983.

[Sharir, 1985] M. Sharir. Intersection and Closest Pair Problems for a Set of Planar Discs. *SIAM Journal of Computing*, 14(2):448–468, 1985.

[Shigley, 1996] Joseph E. Shigley, editor. *Standard Handbook of Machine Design.* McGraw Hill, 2nd edition, 1996.

[Shimada *et al.*, 1989] Kenji Shimada, Masayuki Numao, Hiroshi Masuda, and Shinji Kawabe. Constraint-based Object Description for Product Modeling. In F. Kimura and A. Rolstadås, editors, *Proc. Computer Applications in Production and Engineering CAPE'89*, pages 95–106. International Federation for Information Processing (IFIP), Elsevier Science Publishers B.V., 1989.

[Smith, 1985a] R. Smith. Offline Robot Programming Using Graphics. In *SIGGRAPH '85, Panel: Real-Time Simulation in the Real World.* San Francisco, California, July 26, ACM, 1985.

[Smith, 1985b] R.C. Smith. Fast Robot Collision Detection Using Graphics Hardware. In *Proc. IFAC Symposium on Robot Control (SYROCO '85)*, pages 277–282. Barcelona, Spain, 1985.

[Smith, 1996] B. R. Smith. Virtual Factory. http://cewww.eng.ornl.gov/amnii/projects/vir_fac.html, Oak Ridge Center for Manufacturing Technologies, Oak Ridge, Tennessee, October 1996. [Accessed July 4, 2000].

[Sperlich and Schaermeli, 1997] Tom Sperlich and Gernot Schaermeli. Höhlenbewohner: 3D-Umgebungen der virtuellen Art (Cave Inhabitants: 3D Environments of the Virtual Kind; in German). *c't*, 97(12), 1997.

[Spur and Krause, 1984] G. Spur and F.L. Krause. *CAD-Techniken (Techniques of CAD; in German).* Carl Hanser Verlag, 1984.

[Spur *et al.*, 1976] G. Spur, G. Tannenberg, K.G. Dicke, and W. Weisser. Industrieroboter in der spanenden Fertigung (Industrial Robots Applied to Machining Purposes; in German). *ZwF CIM - Zeitschrift für wirtschaftliche Fertigung und Automatisierung*, 71(1):4–7, 1976.

[Spur *et al.*, 1988a] G. Spur, F.-L. Krause, and M. Lutz. Geometrische Grenzfälle in CAD-Systemen (Geometric Exceptions in CAD-Systems; in German). *ZwF CIM - Zeitschrift für wirtschaftliche Fertigung und Automatisierung*, 83(1):35–39, 1988.

[Spur *et al.*, 1988b] G. Spur, F.-L. Krause, and M. Lutz. Konsistenzsicherung von Datenstrukturen in geometrischen Modellierern (Saving of Consistency within Geometric Modelers; in German). *ZwF CIM - Zeitschrift für wirtschaftliche Fertigung und Automatisierung*, 83(3):128–132, 1988.

[Srinivasan and Basdogan, 1997] M. A. Srinivasan and C. Basdogan. Haptics in Virtual Environments: Taxonomy, Research Status, and Challenges. *Computers and Graphics, Special Issue on Haptic Displays in Virtual Environments*, 21(4):393–404, 1997.

[Stifter, 1986] Sabine Stifter. Intersection of Superellipsoids with Spheres and Planes. Technical report no. 86-11.0 (RISC-Linz Series), RISC-Linz, J. Kepler University, Linz, Austria, Europe, 1986.

[Stifter, 1989] Sabine Stifter. The Roider Method: A Method for Static and Dynamic Collision Detection. Technical report no. 89-40.0 (RISC-Linz Series), RISC-Linz, J. Kepler University, Linz, Austria, Europe, 1989.

[Stobart, 1987] R.K. Stobart. Collision Detection for Offline Programming of Robots. In A. Storr and J.F. McWaters, editors, *Proc. Offline Programming of Industrial Robots*, pages 107–117. Elsevier Science Publishers B.V. (North Holland), 1987.

[Straßer and Wahl, 1995] W. Straßer and F. M. Wahl. *Graphics and Robotics.* Springer, 1995.

[Sun, 1989] Wenhuan Sun. A Study on Representing Geometric Object by Combining the Extended Octrees and Boundary Representation Method. In Tianmin Pan, Bingshu Tong, and Guang Yang, editors, *Proc. 2nd International Conference on CADD and Manufacturing Technology*, pages 199–206, Hangzhou, China, 1989.

[Surakka and Pulkrab, 1997] Esa Surakka and Jürgen Pulkrab. VIFTOO Operating Manual. Technical report, School of Engineering, Pohjois-Savo Polytechnic, Varkaus, Finland, 1997.

[Sutherland, 1965] I.E. Sutherland. SKETCHPAD: A Man-Machine Graphical Communication System. Technical report 296, Lincoln Laboratory, Massachusetts Institute of Technology, USA, 1965.

[Thibault and Naylor, 1987] William C. Thibault and Bruce F. Naylor. Set Operations on Polyhedra Using Binary Space Partitioning Trees. In Maureen C. Stone, editor, *Computer Graphics*, volume 21/4, pages 153–162. ACM SIGGRAPH, 1987.

[Thomas and Rozenblit, 1995] Carsten Thomas and Jerzy W. Rozenblit. Projection-Based Knowledge Representation for Concurrent Engineering). In *Proc. IEEE Intl. Conf. on Systems, Man, and Cybernetics*, pages 3863–3868, Oct. 22–25, 1995, Vancouver, BC, Canada, 1995.

[Tiller, 1983] W. Tiller. Rational B-Splines for Curve and Surface Representation. *IEEE Computer Graphics & Applications*, 3(6):61–69, September 1983.

[Tilove and Requicha, 1980] R.B. Tilove and A.A.G. Requicha. Closure of Boolean Operations on Geometric Entities. *Computer Aided Design*, 12(5):219–220, September 1980.

[Tilove, 1981] R.B. Tilove. Exploiting Spatial and Structural Locality in Geometric Modeling. Technical report no. 38, Production Automation Project, University of Rochester, Rochester, NY 14627, USA, 1981.

[Tomiyama and Yoshikawa, 1990] T. Tomiyama and H. Yoshikawa. Towards Intelligent CAD Systems. In *Proc. ISATA 90, Volume II*, pages 86–93, 1990.

[Toppinen, 1997] Arto Toppinen. Virtual Factory for Education. Technical report, School of Engineering, Pohjois-Savo Polytechnic, Varkaus, Finland, 1997.

[Torguet and Caubet, 1995] Partice Torguet and Ren Caubet. VIPER (Virtuality Programming Environment): A Virtual Reality Applications Design Platform. In *Proc. 2nd Eurographics Workshop on Virtual Environments*, 1995.

[Udupa, 1976] Shriram M. Udupa. *Collision Detection and Avoidance in Computer Controlled Manipulators.* PhD thesis, Electrical Engineering Dept., California Institute of Technology, 1976.

[van Damme, 1993] J. van Damme. *Studie van het Mechanisme van de Halsbeweging bij Chelodina Longicollis (Study of the Mechanism of the Neck Motions of a Chelodina Longicollis; in Dutch).* PhD thesis, Department of Biology, Antwerp University, Belgium, Europe, 1993.

[van Hook, 1986] Tim van Hook. Real-Time Shaded NC Milling Display. In David C. Evans and Russell J. Athay, editors, *Computer Graphics*, volume 20/4, pages 15–20. ACM SIGGRAPH, 1986.

[VDI, Verein Deutscher Ingenieure, 1983] VDI, Verein Deutscher Ingenieure. *Programmieren numerisch gesteuerter Handhabungssysteme – Adressierung von Koordinaten und Funktionen (Programming of Numerically Controlled Handling Devices – Addressing of Coordinates and Functions; in German).* VDI 2864, 1983.

[Vig *et al.*, 1988] Michelle Vig, Kevin Dooley, and Patrick Starr. The Importance of Simulating Cell Activities to Computer Integrated Manufacturing. In *Proc. AUTOFACT'88*, pages 13.11–13.18, Chicago, Illinois, USA, 1988. Society of Manufacturing Engineers, Computer and Automated Systems Association (CASA/SME).

[Vince, 1995] J. Vince. *Virtual Reality Systems.* Addison-Wesley, 1st edition, 1995.

[Voronoi, 1908] G. Voronoi. Nouvelles applications des paramètres continus à la Théorie des Formes Quadratiques. Recherches sur les parallélloèdres primitifs (New Applications of Continuous Parameters in the Theory of Quadratic Forms. Investigations on the Primitive Parallelohedra; in French). *J. reine angew. Math.*, 134:198–287, 1908.

[Vossloh, 1983] M. Vossloh. Breites Spektrum in der Entwicklung numerischer Steuerungen (Huge Variety in the Development of NC Controllers; in German). *Werkstatt und Betrieb*, 116(9):554–559, 1983.

[Wagner, 1989] Gernot P. Wagner. Simulation spart viel Geld (Simulation Saves Much Money; in German). *Megatech*, 5(6):18–19, 1989.

[Ward, 1997] Matt Ward. Advanced Topics in Computer Graphics. Lecture Notes (CS563), Worcester Polytechnic Institute, 100 Institute Road, Worcester, MA 01609-2280, USA, 1997.

[Ware *et al.*, 1993] Colin Ware, Kevin Arthur, and Kellogg S. Booth. Fish Tank Virtual Reality. In *Proceedings of the INTERCHI '93 Conference on Human Factors in Computing Systems*, pages 37–42, Amsterdam, April 1993.

[Warnecke and Altenhein, 1986] H.J. Warnecke and A. Altenhein. Zwei Verfahren zur Kollisionserkennung und -vermeidung bei der Offline–Programmierung von Industrierobotern (Two Methods for Collision Detection and Avoidance in the Field of Industrial Robot Off Line Programming; in German). *Robotersysteme*, 2:163–169, 1986.

[Waterbury, 1983] R. Waterbury. Factory Simulation: Testing Automation's "What Ifs". *Assembly Engineering*, pages 36–39, July 1983.

[Wätzig and Cajar, 1990] R. Wätzig and A. Cajar. CNC für Laserstrahlbearbeitung mit werkstattorientierter Programmierung (CNC Laser Welding using WOP; in German). *Fertigungstechnik und Betrieb*, 40(10):585–587, 1990.

[Weber and Dürr, 1990] H. Weber and H. Dürr. NC-Daten wissensbasiert ermitteln (Determining NC Data Using a Knowledge Base; in German). *ZwF CIM – Zeitschrift für wirtschaftliche Fertigung und Automatisierung*, 85(11):572–575, 1990.

[Webster's, 1995] Webster's. *New Encyclopedic Dictionary.* Black Dog and Leventhal Publishers Inc., 1995.

[Weck and Stöck, 1985] M. Weck and H. P. Stöck. Kollisionsvermeidung bei Industrierobotern (Collision Avoidance for Industrial Robots; in German). *VDI-Z - Entwicklung, Konstruktion, Produktion*, 127(3):71 – 79, 1985.

[Weseslindtner, 1984] H. Weseslindtner. *Industrieroboter — Entwicklung und Herstellung in Österreich (Industrial Robots — Development and Manufacturing in Austria; in German).* TU Vienna, Austria, Europe, 1984.

[Widdoes, 1974] C. Widdoes. A Heuristic Collision Avoider for the Stanford Robot Arm. Technical report Stanford C.S. Memo 227, Stanford University, Stanford, USA, 1974.

[Wilhelms and Skinner, 1990] J. Wilhelms and R. Skinner. A Notion for Interactive Behavioral Animation Control. *IEEE Computer Graphics and Applications*, 90(5):14 – 22, 1990.

[Willim, 1989] Bernd Willim. *Leitfaden der Computer Graphik (Principles of Computer Graphics, in German).* Drei-R-Verlag, Berlin, Germany, 1989.

[Winkowski, 1989] Jozef Winkowski. Formal Theories of Petri Nets and Net Simulation. In *Proceedings of the SIGPLAN '89 Conference on Programming Language Design and Implementation*, 1989.

[Wolf, 1996] Werner Wolf. OASIS - An Object-Oriented Approach to Sensor-Integrated Simulation. Diploma thesis, RISC-Linz, Dept. of Mathematics, Johannes Kepler University, Linz, Austria, 1996.

[Wunsch, 1998] Susi Trautmann Wunsch. *The Adventures of Sojourner : The Mission to Mars That Thrilled the World.* Mikaya Press, 1998.

[Ying and Zhou, 1988] D.N. Ying and M.W. Zhou. SMSZU – A Solid Modeler with a TST Processor. In S.M. Slaby and H. Stachel, editors, 3rd *Int. Conf. Engineering Graphics & Descriptive Geometry*, pages II:312–315, TU Wien, Austria, July 1988.

[Yu and Khalil, 1986] Z. Yu and W. Khalil. Table Look Up for Collision Detection and Safe Operation of Robots. In *Proc. IFAC Theory of Robots*, pages 343–347. Vienna, Austria, 1986.

[Zarrugh, 1985] M.Y. Zarrugh. Display and Inertia Parameters of Superellipsoids as Generalized Constructive Solid Geometry Primitives. In *Proc. of the 1985 ASME International Computers in Engineering Conference and Exhibition (Vol. 1)*, pages 317–328. August 4–8, 1985, Boston, MA, 1985.

[Zeigler, 1984] Bernard P. Zeigler. *Multifacetted Modeling and Discrete Event Simulation.* Academic Press, 1984.

Index